Political Ecology

Tor A. Benjaminsen • Hanne Svarstad

Political Ecology

A Critical Engagement with Global Environmental Issues

palgrave
macmillan

Tor A. Benjaminsen
Department of International Environment and Development Studies
Faculty of Landscape and Society
Norwegian University of Life Sciences
Ås, Norway

Hanne Svarstad
Development Studies, IST – LUI
Faculty of Education and International Studies, OsloMet – Oslo Metropolitan University
Oslo, Norway

ISBN 978-3-030-56035-5 ISBN 978-3-030-56036-2 (eBook)
https://doi.org/10.1007/978-3-030-56036-2

This Palgrave Macmillan imprint is published by the registered company Springer Nature Switzerland AG
The registered company address is: Gewerbestrasse 11, 6330 Cham, Switzerland

Endorsement

'The book describes our common present with unsentimental urgency. Benjaminsen and Svarstad demonstrate the complexity of human engagement with the scarce resources of our planet, and the analytical pathways offered by political ecology. The book's many vivid examples underscore how power is always part of the equation: people + their environment.'

Christian Lund
Professor, University of Copenhagen, Denmark

In loving memory of Anna Benjaminsen, 1936–2020

Preface and Acknowledgements

This book provides an introduction to political ecology, which is an interdisciplinary field of critical studies of environmental issues. We present many case studies and examples from local sites in both the Global South and Global North, linking local environmental processes to the national and global levels. We believe that the book can be used to introduce students to political ecology at both the undergraduate and postgraduate levels, but we also think it may be of interest to a general readership engaged in environmental issues.

Political ecology has grown as an academic field during the last few decades, attracting in particular many young scholars and students from social, human and natural sciences. This is a critical approach examining power in environmental governance, and especially related to injustice and environmental decline. Being critical also means studying knowledge production and the exercise of power within science itself and its links to policy-making. Our intention is to equip students with tools to question what is often taken for granted in environmental policy formulations and in environmental science.

The book presents a number of such examples of examining what tends to be taken for granted. It is based on a textbook in political ecology first published in Norwegian by Universitetsforlaget (Scandinavian University Press) in 2010 and in a revised version in 2017. This English edition is not merely a translation. It is an extensively revised version compared to the last Norwegian edition. All chapters have been reworked, some more than others. While ► Chaps. 2 and 10 are completely new, ► Chaps. 4, 6, 7, 8 and 9 have been updated and rewritten with new cases.

Political ecology has become a broad field including several directions, and we do not pretend to cover all of the current trends. Theoretically, we first of all draw on discourse theory, Marxist political economy, various other social science theories and insights from natural science. Most of the examples and case studies are taken from our own empirical research, mostly in Africa and Norway.

We are grateful to the following colleagues who commented on single chapters or the whole manuscript in Norwegian: Jørund Aasetre, Hans Petter Andersen, Mikael Bergius, Axel Borchgrevink, Halvard Buhaug, Karoline Daugstad, Nils Petter Gleditsch, Ingrid Guldvik, Eirik Jansen, Andrei Marin, Kristen Nordhaug, Knut Nustad, Mariel Aguilar Støen, Nicholas Tyler and Andreas Ytterstad.

Research is a collective effort, especially in an interdisciplinary field such as political ecology, in terms of both building on the previous work of colleagues and carrying out the research together. Most of the examples, case studies and conceptual ideas in this book have been studied or developed in collaboration with colleagues who we have enjoyed working with. We therefore acknowledge with appreciation the contributions by Jens Aune, Boubacar Ba, Mikael Bergius, Ian Bryceson, Connor Cavanagh, Inger Marie Gaup Eira, Pierre Hiernaux, Kathrine Johnsen, Thembela Kepe, Faustin Maganga, Mikkel Nils Sara, Sayuni Mariki,

Ragnhild Overå, Jon Pedersen, Nitin Rai, Rick Rohde, Espen Sjaastad, Silje Skuland, David Tumusiime, Mats Widgren and Poul Wisborg.

In addition, we are grateful to other colleagues who have commented on various draft papers that have been integrated in this book in different ways and who in meetings and discussions helped us formulate some of the ideas we present here. In particular, we would like to thank Dan Brockington, Jill T. Buseth, Frances Cleaver, Denis Gautier, Jens Friis Lund, Synne Movik, Christine Noe, Paul Robbins, Jesse Ribot and Teklehaymanot Weldemichel.

We are indebted to Per Robstad at Universitetsforlaget who was instrumental in facilitating the Norwegian editions. Rachael Ballard and Joanna O'Neill at Palgrave Macmillan have been helpful throughout the process from the first lunch meeting to discuss ideas to patiently guide us through the final details of the manuscript format.

We also thank our daughter Eline Benjaminsen, who is a photographer. She critically engaged with our use of photos and had clear ideas about what to use and not to use. She also took the photo on the cover, which is a product of photogrammetry from her project 'Footprints in the Valley'. In addition, she helped us keep track of all the references.

Over the years, we have also enjoyed and benefitted from numerous discussions and exchanges of ideas with students at our universities—the Norwegian University of Life Sciences and Oslo Metropolitan University. Teaching engaged students was one of the main inspirations for writing this textbook.

We are grateful to all the scholars who contribute to the growing field of political ecology, and we hope that this book will inspire new readers and contributors. We all find much inspirational energy within the international political ecology network POLLEN (► https://politicalecologynetwork.org/welcome/). This is a network not only with many impressive academics, but also with great collegiality, a friendly but also critical (!) atmosphere, and with shared commitments to solidarity, social justice and environmental sustainability.

Last but not least, we are thankful to all the farmers and pastoralists, and other experts on livelihoods and environmental change in various rural settings that we have learned from over many years. Without their knowledge and insights, we would not have been able to write this book.

We both are responsible for the book at large and all chapters. Parts of the book build on research we have conducted together, but much of it is based on research each of us has conducted separately. Hanne has written most of ► Chap. 1, while Tor has written most of ► Chaps. 2 and 4. ► Chaps. 3, 5 and 6 are mainly written by Hanne, while ► Chaps. 7, 8, 9 and 10 are mainly written by Tor.

Tor A. Benjaminsen
Oslo, Norway

Hanne Svarstad
Oslo, Norway
August 2020

Contents

List of Photos

List of Figures

List of Tables

Political Ecology on Pandora

Contents

T. A. Benjaminsen, H. Svarstad, *Political Ecology*, https://doi.org/10.1007/978-3-030-56036-2_1

Trailer

The Canadian filmmaker James Cameron released his first Avatar movie in 2009. It takes place on the moon Pandora, far away from the solar system of the Earth, and 150 years into the future. A mining company from Earth has established a base from which it extracts a mineral called *unobtanium*. The mineral is transported to Earth and used in energy production. It is indeed a profitable operation. At Pandora, there are Na'vi, a human-like people. The Na'vi society has features resembling many indigenous peoples today. They live in a close relationship with their planet's wild, beautiful and bountiful nature. The richest deposit of unobtanium lies beneath the home site of one of the Na'vi clans. Either through the use of negotiation or force, the mining company therefore intends to remove the clan and destroy the land upon which the Na'vi depend. The company has hired a small group of scientists to carry out a participatory observation of the Na'vi in order to find out what to do with them. This intention is hidden for the Na'vi. At the same time, the company employs a large military force led by a colonel who is eager to 'solve the problem' by using force. The film begins when the marine veteran, Jake Sully, arrives on Pandora to join the research group.

In the beginning, Jake is not at all accepted by the Na'vi, but his relationship with them improves. The film has a romantic element when Jake falls in love with the Na'vi princess Neytiri. When the colonel starts his armed attack on the Na'vi, it is Jake who heads up the resistance. This seems to be an easy battle for the Colonel, with his well-trained troops and high-tech military equipment against the Na'vi, who appear to be a weak and 'primitive' people with bows and arrows. But David was able to outwit Goliath, and in movies everything is possible. In Avatar, Cameron has drawn his inspiration from film genres such as science fiction and war films. The parallels to western films are also obvious, as this film reminds us of the fights between cowboys and American Indians.

You may now start to wonder: Why do we begin this book with a science fiction movie from outer space? Some may even argue that this is total nonsense, and there is no connection whatsoever between political ecology and the film Avatar. Truly, political ecology is a field of study and a research approach aimed at nonfiction studies of human life in their environments on the real planet Earth. And as we know today, there is no Planet B.

Nevertheless, we use the fiction movie about the struggle at Pandora as a 'case' to introduce what we argue to constitute key building blocks of political ecology. Despite being fiction, Avatar embraces a theme that political ecology often highlights. It is about conflicts where local and sometimes indigenous communities are approached by external actors who try to appropriate their land and natural resources.

Survival International (SI) is an organization for indigenous peoples. In a press release when the first Avatar movie was launched, SI stated that '*Avatar* is real'. SI's director, Stephen Corry, said: 'Like the Na'vi of "Avatar", the world's last-remaining tribal peoples – from the Amazon to Siberia – are also at risk of extinction, as their lands are appropriated by powerful forces for profit-making reasons such as colonization, logging and mining.' (Survival International 2010).

A man of the Penan people in Sarawak, Malaysia, made the following comment to SI: 'The Na'vi people in "Avatar" cry because their forest is destroyed. It's the same with the Penan. Logging companies are chopping down our big trees and polluting our rivers, and the animals we hunt are dying' (Survival International 2010).

The San people, or Bushmen as they call themselves in Botswana, are a hunter-gatherer people who live in the dry areas of southern Africa. Jumanda Gakelebone, is a spokesperson for the organization *First People of the Kalahari*. He said: 'We the Bushmen are the first inhabitants in southern Africa. We are being denied rights to our land and appeal to the world to help us. "Avatar" makes me happy as it shows the world about what it is to be a Bushman, and what our land is to us. Land and Bushmen are the same.' (Survival International 2010).

Mikkel Berg-Nordlie is a researcher and a Norwegian Sámi who works on questions about indigenous peoples and ethnic minorities. He emphasizes that Pandora and Norwegian Sápmi are two completely different places. Yet, he believes that the Sámi people can recognize themselves in the Na'vi: 'We have also experienced discrimination and forced intervention in our areas that have been to our disadvantage'.

The release of the first Avatar movie sparked off extensive web debates among academics. Some critiques pointed out that the hero of the film is white. Blogger Annalee Newitz (2009) argued that Avatar belongs to a category of films based on 'white guilt fantasy':

> It's not just a wish to be absolved of the crimes whites have committed against people of color; it's not just a wish to join the side of moral justice in battle. It's a wish to lead people of color from the inside rather than from the (oppressive, white) outside.

Moreover, the depiction of the Na'vi has been criticized for being based on a stereotype that white people in the West have of indigenous peoples. Van der Horst (2010) argues that the Na'vi constitute a reversed mirror image of Americans: 'The more rational, technological, calculative and materialist the Americans/Westerners are, the more irrational, spontaneous, mystical and backward the Na'vi are.'

This discussion also relates to recent debates about 'decolonization' in general and of decolonization of environmental conservation in particular (Adams and Mulligan 2003; Martin et al. 2016), which is about whose worldviews, values, and knowledge are taken into account in academic, artistic and journalistic presentations as well as in practical politics related to environmental conservation. Or put differently, the discussion about decolonization is for instance about on whose terms stories are told. In the case of Avatar, this is a contested question.

What Political Ecologists Study

Scholars in the field of political ecology often start out their studies at the local level. Early in the history of political ecology, the focus was mainly on people and environments in rural areas of the Global South. In the last decade or two, however, political ecology has also moved North, as well as included the study of urban issues. In this book, we emphasise the importance of political ecology case studies that are conducted in the Global North and South and reveal ways that spaces in different parts of the world are interconnected and influenced by powerful actors of both the North and the South.

Three different processes are frequently studied in political ecology. The first is when international capital invests in a business in a way that threatens local livelihoods through environmental impacts and access to land. This might involve mining as in the case of Avatar, or it might be about agricultural production, manufacturing or the construction of power plants, or in the case of urban political ecology, when poor or marginalized communities are squeezed out of a neighbourhood following gentrification.

Second, external actors such as big international conservation organizations may influence national governments, usually in the Global South, to restrict local use of land and natural resources in order to protect the global environment and climate. This includes initiatives to establish new national parks, or climate mitigation projects such as production of biofuel or to sequester carbon to conserve forests or establish new forest plantations.

Third, some political ecologists also study environmental change: What processes of environmental change do actually take place? What are impacts as well as causes. Examples are desertification, deforestation and soil erosion. Often, political ecology has questioned the general scope of these processes, their root causes as well as dominating views and policies related to these environmental issues.

Political ecology is a relatively new academic tradition that emerged in the 1980s, which provides perspectives and analytic tools to investigate the three mentioned processes among others. Local situations are studied in the light of national and global influences, and political ecology provides a critical alternative to other ways of studying environmental issues. This implies that political ecologists examine power relationships and question mainstream claims about environment and development that often are taken for granted.

There are two common misunderstandings about political ecology. The first is that this is a school of thought within the discipline of ecology, while political ecology is actually an interdisciplinary field. Often elements from ecology and other natural sciences play important roles, but political ecology draws primarily on social sciences. Therefore, 'ecology' should be seen to reflect the field's broad focus on environmental issues. Some scholars with background from natural sciences as well as from the humanities do political ecology studies, although the field is dominated by scholars from various social sciences, first of all human geography and anthropology, but also development studies, sociology and political science.

The second misunderstanding is that political ecology is political in the sense that political statements are made on environmental questions without sound foundations in empirical knowledge. Thus, some believe that political ecology merely implies claims-making on environmental issues without empirical research. However, as a research-based approach, the ambition for scholars within political ecology, as in any other field, should be to be open to any findings based on empirical investigation.

As we will soon show, the 'political' in political ecology is actually drawn from the older field of *political economy*. In addition, and similar to fields such as political science, political geography and political philosophy, the political aspect of political ecology involves a focus on various forms of power.

In this book, we introduce political ecology to students who are interested in issues of environment and development, but who are not familiar with this scholarly field, its approach and current debates within this field. We do this by introducing theoretical concepts and theories combined with a series of empirical local case studies mostly drawn from our own research. In these case studies, we try to show how 'the global' can be studied in 'the local'.

The case studies are both from the Global South (mostly from Africa) and the Global North (mostly from our own country Norway). These geographical levels are not studied in isolation, but are closely inter-connected. This is for instance the case with how Northern pharmaceutical companies use biodiversity and local knowledge in Tropical countries for bioprospecting of new medicines and how Northern activists relate to this activity (▶ Chap. 3); how Norway as an oil-producing nation has become a global leader in promoting forest conservation in the Global South to mitigate climate change (▶ Chap. 6); and how international capital has shown a relatively recent interest in accessing land for large-scale agricultural investments or forest plantations in Africa in particular (▶ Chap. 9).

In this chapter, we introduce some perspectives and research tools that are central in political ecology. ▶ Chapter 2 deals with the emergence of political ecology and the background to different theories and perspectives that are used in the field including some internal directions and debates. In other words, both ▶ Chaps. 1 and 2 are theoretical chapters. In ▶ Chaps. 3, 4, 5, 6, 7, 8, and 9 we highlight particular themes or theoretical discussions in the field based on the presentation of various case studies. Finally, in ▶ Chap. 10 we synthesize the experiences and insights that our approach provides at the theoretical level, and we present some thoughts about possible ways forward for political ecology.

New syntheses and continuous integration of new theories are a hallmark of political ecology. An important aim is to help to understand questions about environment and society, and to do so, to produce knowledge from a broad scope of social as well as other theories. In the following, we present what we see as ten syntheses that may help a new reader not familiar with the field to understand what political ecology is really about. For undergraduate students and students from other fields than social sciences, the rest of this chapter also provides a basic introduction to some key building blocks and discussions of social sciences as such.

1.1 First Synthesis: Social and Natural Sciences

As mentioned, political ecology is an approach that is dominated by social sciences, but aims to build a bridge to the natural sciences. The ecology focus of political ecology has not been drawn directly from the natural science discipline of ecology, but instead more indirectly builds on the (mostly) social science approaches of human and cultural ecology where social groups are studied as part of ecosystems (see ▶ Chap. 2).

Nevertheless, some political ecologists work closely with natural scientists. Others specialize on a thematic field and are able to read and critically assess the scientific literature in that field as well as bringing these assessments into their own analyses.

Natural sciences can contribute to the establishment of an understanding of, for instance, the spatial and ecological aspects of environmental governance or conflicts that are discussed. In a lot of political ecology research, such aspects may have a relatively modest role and are limited to providing an overview of the knowledge that natural scientists have on the subject. At the same time, political ecology involves a critical assessment of the premises for all scientific knowledge. In a number of political ecology projects, it is essential to possess a good deal of knowledge of how various factors play out spatially and it may be necessary to involve methods such as remote sensing, as well as soil and vegetation analysis.

Some political ecology aims to combine natural science studies with knowledge obtained through qualitative methods to identify what indigenous or local knowledge says about the same environmental processes. Such combination of knowledge has been called *hybrid science* (Batterbury et al. 1997; Forsyth 2003).

In Avatar, Jake Sully's research group uses a combination of social and natural sciences. On the one hand, they conduct an ethnographic study of the perspectives and way of living of the Na'vi. On the other hand, they research Pandora's diverse flora and fauna. In these nature studies, the scientists take as a starting point the Na'vi's knowledge about nature. They even take participant observation to a new level by entering into Na'vi bodies (avatars). In ◘ Photo 1.1, there is a Na'vi gathering discussing whether or not the Na'vi should meet the mining company's destruction of the forest with violent resistance. The research leader Grace Augustine (Sigourney Weaver) speaks out against such resistance, arguing that it will only make their situation worse.

◘ **Photo 1.1** Are the scientists in the Avatar movie political ecologists? (© *20TH CENTURY FOX*)

The researchers also carry out quantitative investigations of nature. This is, for instance, apparent when Grace Augustine is mortally wounded and the Na'vi carry her to their legendary 'soul tree'. Grace realizes where she is, forgets her own serious situation and exclaims as an engrossed scientist: 'I need to take some samples!'

Nevertheless, the research team's combined efforts of acquiring knowledge about both culture and nature is not sufficient to consider this research as political ecology. Focus on power relations, and the researchers' role, plays a key role in political ecology. At the beginning of the film, we see that the scientists without further reflection allow themselves to be used as part of the mining company's colonization of Pandora, while they towards the end take an active role in support of the Na'vi.

1.2 Second Synthesis: Three Aspects of Environmental Governance

Political ecology is about how the natural environment is managed by people. It deals with landscapes that appear as little influenced by humans, as well as cultural landscapes where the human impact is more apparent. As a consequence of environmental governance, people are able to eat, build houses, produce commodities, and so on. At the same time, many of us value the existence of a broad variety of animals and plants (biodiversity).

Humans depend on their environment to survive. What is done to nature today may have serious consequences for the resources nature will provide for people in the future. However, the environment is not managed in such a way today that all people can have a safe life materially, since about one billion live in extreme poverty.

We can divide environmental governance into three aspects: use, conservation and distribution (◘ Fig. 1.1). Human *use* of nature constitutes a prerequisite for satisfying needs and wants. *Conservation* implies measures implemented to maintain certain natural conditions for the future. *Distribution* implies that different individuals and groups are affected differently in terms of economic gains and losses related to uses of nature. Likewise, specific instances of conservation have various distribution consequences for different people.

Historically, economic development from the industrial revolution brought about an increasingly extensive use of nature. Large social disparities resulted in a focus on distribution in terms of class struggle from the mid-nineteenth century in Europe. Conservation of nature is a theme that first really received political significance from the growth of the environmental movement in the 1960s. ► Chapter 4 in particular includes discussions of distributional injustice in cases of nature conservation.

Specific situations can also be considered in light of the three aspects of environmental governance. Let us again look at *Avatar* as an example. The movie is about a conflict between two actor groups that wish to use Pandora's natural resources in different ways. The Na'vi want to continue their traditional natural resource use, and the film suggests that their use of nature is in consistence with

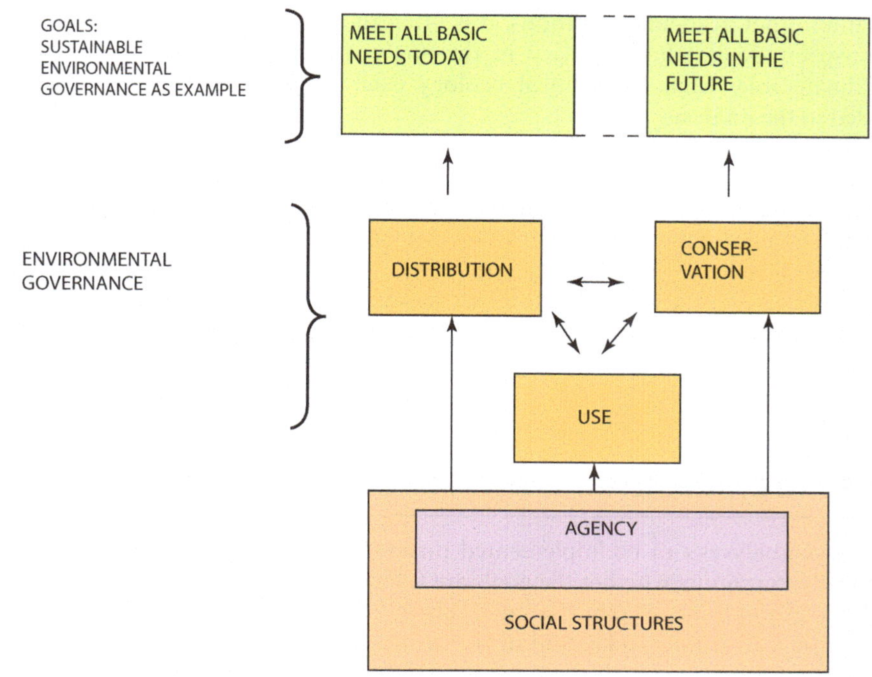

Fig. 1.1 The three aspects of environmental governance (use, conservation, distribution) are produced through agency (acts by actors). Social structures limit and facilitate these actions. Normative analyses can be undertaken by defining specific goals (such as sustainable environmental governance) and by empirically examining the extent to which they are fulfilled in specific cases. (Source: Created by the authors)

long-term conservation of nature. The mining company arriving at Pandora introduces a completely different view of nature and how to use it. The company is not interested in the forest or fauna, but in extracting the mineral unobtanium to sell it on Earth. The mining operation involves the destruction of nature and constitutes a threat to what turns out to be a sophisticated natural system. Therefore, this form of land use is also a threat to the very existence of the Na'vi themselves. The distributional aspects of costs and benefits are that the mining company is likely to achieve a huge profit from the sale of the mineral, while the Na'vi face the destruction of their livelihoods. At the beginning of the film, we see how the company wants to negotiate with the Na'vi, but it is clear that the company is willing to pay the Na'vi very little compared to the profits that it expects from selling the mineral.

Within political ecology, special cases of environmental governance can be investigated and analysed in relation to each of the three aspects as well as their combinations. Such research requires knowledge about the social science dimensions of each of the aspects. At the same time, it is also important to have relevant knowledge about natural dimensions of the species and areas that will be made subject to use and conservation.

Within other academic traditions, it is often the case that one or more of the three aspects do not have a place in the analysis at all – or that they play a subordinate role, while in political ecology each of these aspects are usually included in the analysis.

Students who conduct a political ecology study of a case may start out acquiring knowledge to describe interesting features concerning each of the three aspects of environmental governance. Thereafter, they may move upwards in ◘ Fig. 1.1 and conduct a normative analysis of environmental governance. This implies comparing the observed situation with specific norms or ideals. Students may also move downward in the figure to examine factors that explain the situation. As we will show further in this chapter, actors and social structures here constitute a central conceptual combination, and it is important to examine forms of power that actors exercise.

1.3 Third Synthesis: Normative and Empirical Analyses

Normative analyses can be implemented through the specification of goals, and empirically examining whether the goals are fulfilled and to what degree. This goal could for example be sustainable environmental governance (as in ◘ Fig. 1.1) meaning that all people today and in the future get their basic needs met. These goals are in compliance with the most used definition of *sustainable development* from the Brundtland Commission's report *'Our Common Future'*: a development that 'meets the needs of the present without compromising the ability of future generations to meet their own needs' (The World Commission on Environment and Development 1987).

Meeting these two goals implies first that natural resources today are *used* in a manner that ensures the production of basic needs for all people, such as food and medicines. Secondly, the use must not occur at the expense of *conservation* of nature's future ability to provide resources for such production. Thirdly, the income and expenses from the current use and conservation must be *distributed* in a manner that make poor and disadvantaged people able to meet their basic needs.

In the fictional case of Pandora, it is easy to conclude that the mining operation is in contrast to both goals, because it will imply a fast and irreversible degradation of nature and thereby threaten the Na'vi's livelihoods immediately and permanently.

Ever since the release of the Brundtland report, the notion of sustainable development has been used in very different ways and often as a vague idea of development that is based on economic growth and the present market-based economic system (also called *capitalism*), and argued to improve both social conditions and the environment. Since 2015, the Sustainable Development Goals (SDG) of the United Nations provide a set of 17 goals to be implemented in each country by 2030. The implementation of these goals involves a range of actors such as governments, Non-Governmental Organizations (NGOs) and private companies (United Nations 2015). This approach may, however, be criticized for not taking into account the structural changes necessary to achieve the two ultimate goals in ◘ Fig. 1.1.

Another normative focus in political ecology is to look at the degree of participation and influence among local actors where external actors wish to implement changes. Students who examine participation can for example take as a starting point a scale of participation with a number of different levels ranging from 'manipulative participation' to 'self-mobilization' (Pretty 1995). At the beginning of the Avatar movie, it appears that the mining company wishes to reach a negotiated resolution with the Na'vi. This does not mean that the company is willing to abandon its primary goal of extracting the mineral 'unobtanium' using methods that will destroy Pandora's land. Instead, it is rather an attempt at 'manipulative participation' (see more about participation in ► Chaps. 3, 4 and 5).

Within environmental research, one often encounters attitudes that research should be 'neutral' or 'objective'. However, closer investigation can reveal how specific norms and values are embedded, and, for example, implies that preservation of nature is prioritized, while considerations for people's livelihoods are neglected (Svarstad et al. 2008). Political ecology here builds on the critique of positivism within social sciences and philosophy. Natural as well as social scientists may present their own work as 'neutral' or 'objective', although critical examinations reveal that they are based on specific norms and values with considerable impact on the research conclusions. It is also problematic when this type of assumed 'neutral' research forms the basis for decision-making on environmental governance.

Environmental sustainability, justice and human rights are amongst the norms often subscribed to in political ecology. These norms are broadly shared and are not very controversial, at least not on paper. Nevertheless, they may be taken more seriously in political ecology than in some other scholarly traditions.

1.4 Fourth Synthesis: Agency and Structure

Environmental governance can be seen as a direct result of agency—in other words what actors do. Social structures limit as well as facilitate such actions. When investigating causes of an environmental problem in a specific case, it is useful to try and identify the actors as well as the social structures involved in shaping a particular way of thinking about the environment and how the problems are addressed in practice. Actors and social structures are key concepts in the social sciences in general.

Actors may be individuals or large entities such as organizations, businesses or governmental bureaucracies. They may have really clear or rather hazy goals and values.

There are different ways of conceptualizing social structures. In this book, we use the term about elements that enable or restrict the agency of actors. Social structures include social norms. These are formal or informal rules of behaviour. Furthermore, social structures also entail institutions, which are formalized frameworks for action, such as laws and policies. Discourses constitute yet another type of social structures, and we define them as shared ways of understanding and presenting a particular issue (see ► Chap. 3).

Last but not least, there are also economic structures. Within development studies, the dependency school represents a structure-oriented perspective in its explanation of why some countries are rich and others are poor. André Gunder Frank, Samir Amin, Immanuel Wallerstein and others have shown that development in Western industrialized countries (centre) and former colonies (periphery) are bound together in relationships where raw materials and resources are transferred from the periphery to the centre, and where these centre-periphery connections form the basis for industrial production and economic development. Moreover, there is a social structure with alliances between the elites in the periphery and the centre. According to dependency theory, these aspects of the world system have contributed to development in the centre and underdevelopment in the periphery (Frank 1979; Amin 1977).

More recently, globalization opponents follow in the wake of such a structure-oriented tradition, where they for instance point out negative consequences for countries in the Global South of economic structures at the global level. On environmental issues, political ecologists study how structures of various kinds at the global level can impact on local environments and their use. This may involve economic structures as well as discourses among powerful actors.

Norman Long is one of the best-known contributors to an actor-oriented approach in development studies. He believes that structure-oriented development theories often are too deterministic in that they view social change as only brought about by external influences (Long 1992, 2001). Even where the structural framework is relatively constant, Long believes that actors have more options than what is often assumed. Yet, he does not rule out the importance of social structures. Most environment and development scholars - as well as other social scientists - today realize the necessity of combining structural and actor-oriented explanations in one way or another in their analyses. Nevertheless, there is wide variation in how different scholars place their emphasis, so that some, like Long, are mainly actor-oriented, while others are mainly structure-oriented.

Several social scientists have during recent years established different theoretical syntheses involving a due consideration of both actors and structures. For example, the sociologist, Anthony Giddens, has developed what he calls *structuration theory*. As actors, Giddens focuses on individuals, and social structures encompass rules and resources that the actors may use. Giddens emphasizes that social structures not only restrict actions, but also make actions possible. Actors utilize structures when they act. Furthermore, actions have effects back on the underlying structures, which subsequently change over time. In other words, structure has a *duality* of creating a basis for action and being the results of action (Giddens 1984; Stones 2005).

Actors on Pandora

In *Avatar*, we can distinguish between different actor groups, such as the Na'vi, the mining company, the military force and the research unit. We can also see how different individuals attempt to act in accordance with what they regard as their own interests. Jake Sully's legs are paralyzed. The Colonel gives him the promise that if he does his job in the way the Colonel wishes, Jake can return home to Earth and have an operation paid, and this will make him able to walk again.

The Na'vi can be considered as an actor group, but it is also possible to identify various actors within this community. For example, there are different clans that oppose each other, until Jake comes and unites them in a common struggle against the external enemy (which has made some see the movie as an example of white saviourism). Furthermore, this is a hierarchical society, where the clan's leadership is inherited. The Nature Goddess Eywa is also an actor, who contributes vitally to what is happening on Pandora. So, nature is an actor too, as in Science and Technology Studies (STS) that have also influenced parts of political ecology (Robbins 2007; Goldman et al. 2011). Jake Sully moves back and forth between the mining colony's and the Na'vi's quite different sets of structures, before he eventually chooses sides and fights against the mining colony together with the Na'vi.

1.5 Fifth Synthesis: Realism and Social Constructivism

'Realism' and 'constructivism' are two main theories of science that often are considered to be incompatible. In its purest form, *realism* involves a belief that scientists can describe reality independently of their own norms and values. This is the approach mostly followed in natural science as well as in quantitative, or positivist, forms of social science.

On the other side, *constructivism* implies a view that reality, or rather how we understand reality, is 'constructed' by people who observe and think about it. In this perspective, there are several parallel views of reality, and in the more sterling versions of constructivism, all views (or discourses) are considered equally valid. Note that some scholars use the term 'constructivism', while others write 'constructionism'. Moreover, the word 'social' is often added in the front to emphasize that the constructions are made by contributions of many people; they are in other words social products.

Research rooted in realism focuses on natural or social phenomena, while research based on constructivism is concerned with different views and claims about these phenomena. Political ecology studies of environmental governance often focus on practices and consequences of practices – in nature and for people.

In environmental studies, there is often a main division between research based on realism and constructivism. While constructivists focus on social constructions (i.e. common or competing perceptions of an environmental problem), realists are concerned with studying the problem in its various aspects without dealing with its social constructions. In many political ecology studies, however, scholars try to bridge these two approaches by linking empirical investigations of a phenomenon with social constructions of it.

This middle-position is called 'critical realism' (Forsyth 2003; Neumann 2005) or 'soft constructionism' (Robbins 2020a). It recognizes that there are many aspects of reality independent of our knowledge. This approach accepts that there are different social constructions that impact on how the reality is presented in politics, bureaucracies, media, the public at large, and even in science. However, all these constructions are not seen as being equally important or correct. Instead, claims about reality become the subject of empirical investigations. In other words, 'reality checks' are carried out. An example of such a reality check that we discuss in this book is the idea of 'desertification' in the Sahel. Another example is investigation into whether the rhetoric about participation in conservation actually provides a good description of people's opportunities to influence decision-making in specific cases.

As critical realists or soft constructionists within political ecology, we study how social and environmental practices are characterized by powerful actors, but in addition, we are interested in distinguishing true from false in debates about environmental change. The ideas that powerful actors hold about environmental change may for instance be compared with the empirical investigations of this change.

Some researchers with a scientific foundation in realism often tend to present their research as neutral and objective, although a closer look may reveal how their presentations are based on specific values. Obviously, political ecologists, as researchers in general need to be self-reflective about their own knowledge production and how it may be influenced by specific values or believes. The double position of a critical realist does not provide any guarantee that the research is independent, but it provides a position from which to critically examine how specific believes and views may influence the research.

Social constructions are often studied in terms of discourses and narratives. As the basic definitions, we see *discourses* as ways of viewing specific topics, while *narratives* are stories about particular cases. In ► Chap. 3 we outline a broader conceptual framework on discourses and narratives, and in this and later chapters we present examples of such social constructions and compare them with findings from our own studies of various environmental topics.

Imagine Realism and Social Constructivism on Pandora

Imagine that a research team of political ecologists arrives on Pandora along with Jake Sully. Political ecology does not contain a complete recipe on how a study should be carried out. Nevertheless, we would imagine that the first thing political ecologists would tend to do, would be to examine the social constructions created by the different groups of actors. It would be crucial to reveal the discourse adhered to by powerful actors such as the leaders of the colony. Thereafter, it would be useful to investigate the discourse of the Na'vi, as well as that of the members of the research team led by Grace Augustine. Furthermore, we can imagine that political ecologists would attempt to examine the species and ecology on Pandora, the way of living and livelihoods of the Na'vi, and the consequences that the human colonization might have on the Na'vi and their environment. Thus, such a study would be anchored in both constructivist and realist elements.

1.6 Sixth Synthesis: Different Types of Power

Power is a central theme in political ecology. For example, Robbins (2004:12) defines political ecology as 'empirical, research-based explorations to explain linkages in the condition and change of social/environmental systems, with explicit consideration of relations of power'. Likewise, Paulson et al. (2005:28) write that 'the political' in political ecology is 'used to designate the practices and processes through which power, in its multiple forms, is wielded and negotiated.'

Much political ecology research is carried out as studies of specific conflicts over use of land and natural resources. Investigations are made of causes of the conflicts, and how they evolve. In analyses of power, the researchers apply various combinations of three main perspectives on power. These are actor-oriented power theories and two types of structure oriented theories. We argue that in order to understand power in specific cases, elements from all these three perspectives must be examined (Svarstad et al. 2018).

In studies of conflicts, it is often useful to start with an actor-oriented focus analysing how the different actors involved try to influence environmental governance and establish to what extent various actors succeed and reasons for successes and failures. Some actors may be local, while others may be located in the capital of the country or other distant places.

In the social sciences, there are many actor-oriented power theories. For instance, Russel (1938) defined power as the ability to produce intended effects. This may sometimes be possible to do more or less independently of other actors. If you are hungry and thirsty and live in an affluent society, most people are able to produce intended effects by grabbing a slice of bread or a glass of water without directly involving anybody else. But power does usually involve others.

For example, Max Weber provided a classic contribution to actor-oriented power theories. He defined power as the ability of individuals to realize their will despite resistance from others (Weber 1964). Robert Dahl (1957) gives an example of actor-oriented power theory in which actor A exercises power over actor B, when A can get B to do something that B would not otherwise do. An extreme version of such power arises when someone is forced to do something, which they are strongly against. Such exercise of power is to be found when powerful actors force people to move from their home place in connection with the establishment of a new protected area - or because of a mining operation, as in *Avatar*.

In the actor-oriented power perspective presented by the Norwegian sociologist, Fredrik Engelstad, a 'strong' concept of power is at the same time intentional, relational and result-orientated. *Intentionality* implies that power is exercised through the actions undertaken by someone to achieve something. *Relationality* means that it is about action that happens in the relations between two or more actors. *Result-orientation* (causality) implies that the actions performed have a desired effect. Cases where only two of the three dimensions of power are present, Engelstad also sees as involving power, but of weaker kinds (Engelstad 1999).

We would argue that all the three dimensions should be present to make it meaningful to use the term power. Nevertheless, we would not require that an actor gets a result in total coherence with intentions in order to consider it an exercise of power. In most situations, several actors are involved in trying to obtain results of different kinds, but they succeed to varying degrees.

Both the second and third main power perspectives in political ecology are structure-oriented. As outlined in the previous subchapter, social structures delimit actors' possible actions, and at the same time, they make action possible.

The second perspective is about economic structures and grounded in a Marxist tradition of political economy (see ► Chap. 2), with emphasize on power in the global capitalistic system. Capitalists have power first of all because they own means of production. Workers may obtain power through organizing themselves in trade unions, and thereby conduct negotiations together and chose to strike in order to improve working conditions and salary. However, since production takes place in different parts of the world, there are many hindrances against such organization. If workers lay down work in a mine owned by an international mining corporation, the company might close down the mine and instead conduct their activities in countries without strong trade unions.

In specific conflicts on natural resources, economic structures will constitute important frameworks for what is possible and impossible to do for various groups of actors. Often local conflicts are tied to larger questions about the organization of the economy, where some of the involved groups engage in attempts to influence these structures. A company may, for example, try to influence a national state and politicians to change laws in order for them to usurp land for establishing plantations.

The third main perspective on power is about social structures, and this perspective is associated with the constructivist tradition. This is power exercised by getting others to adopt a way of thinking that is favourable for what they consider 'their own interests'. The sociologist Steven Lukes (2005) provides a variation of such a power theory in what he calls a three-dimensional power perspective. An example of the third dimension is when one actor gets other actors to do something they would otherwise not do, by influencing their wishes. For a government, for instance, this could take place through control of information via media and education, so that people could only get access to presentations of issues and cases decided by the government.

Antonio Gramsci and Michel Foucault are both important sources of inspiration for power perspectives in political ecology. Gramsci (1932/1975) was an Italian Marxist who in the 1920s and 1930s was a prisoner in Mussolini's fascist state. In his prisoner's notes, Gramsci wrote about how ideas and world views by workers and other ordinary people often were in line with the interests of the rulers. For most people, the rulers' perspectives seemed to be the only possible ways of perceiving reality.

Foucault has provided ideas about how the production of perceptions on reality among rulers and the state apparatus influence reality. He has distinguished between four ways of governing – *governmentalities* – ways that state authorities in modernity manage populations and get people to act as the government wishes (Foucault 2008; Fletcher 2010) – see definition on p. 44.

A number of studies of environmental issues pointing out such mechanisms have been carried out (Agrawal 2005; Flyvbjerg 1998; Fletcher 2010). Arun Agrawal first introduced Foucault's power perspective into political ecology through the concept *environmentality*.

Power theories focusing on social structures may be divided into two groups. The first is about structural characteristics that some establish or use in order to obtain particular results. This is exercise of power conducted by intentional actors. In the other group there are theories of social structures that in certain situations get people to think and act in particular ways without this being an intention from anybody. Such non-intentional effects are produced, for example, when many people go by car every day to work. Climate change and local pollution may in this way be non-intentional effects. Neither governments nor automobile companies nor car drivers are likely to have any wishes to contribute to climate change and pollution. With structural changes such as restrictions on cars or improvements of public transport, the effects could be reduced. Intentional exercise of power and non-intentional effects are both social elements with high importance. Nevertheless, the analyses become vague and unclear if their differences are not specified conceptually. We argue that it is important to distinguish between exercises of power on the one hand, and other social aspects such as non-intentional effects of social structures. With other words, these are differences between power and non-intentional operations of 'force'.

Power Resources

In studies of environmental conflicts, we think it is useful to establish the power resources that are available and used by different actors. We will mention nine different types, but this is not an exhaustive list. The specification of power resources implies a focus on actors. At the same time, each of these resources is also related to structure-oriented power perspectives.

1. Power resources might consist of economic resources, such as finance capital (money) or ownership of businesses. Those in possession of such economic assets often have great influence on economic developments. Private owners of large businesses, as well as governmental and non-governmental aid organizations, possess power resources of this type.
2. Ownership, control over and use rights to land and natural resources is a variation of an economic power resource that is particularly important in environmental governance. An example is what is labelled 'land grabbing' or 'green grabbing', where actors, being foreign or national, acquire land, while local smallholders are squeezed out. The land is then used to produce food or biofuels, set aside for climate mitigation, or reserved for conservation and tourism (see ► Chaps. 4, 6 and 9).
3. Power may be exercised on the basis of political resources in the form of influence over policies, laws and public budgets. In all countries were governmental

leaders are elected, the ability to get support from voters provides a type of political power resource.

4. Influence over bureaucracies constitutes a power resource that may be used by elected politicians. Other actors, for example companies, can also exercise great influence on parts of the government apparatus. In some cases, government officials exercise power in ways where they first of all take care of their private economic interests in terms of rent-seeking or blatant corruption.
5. Discursive resources consist of abilities to present specific issues in ways that meet approval by other actors. Several of the chapters in this book show examples of how some actors manage to get wide acceptance for certain views even though research reveals that the claims are not supported by empirical knowledge.
6. There is much wisdom in the proverb that 'knowledge is power'. Production of and access to information constitutes an important power resource, which is often unevenly distributed between the parties in a conflict over environmental governance. One example is information about the increased value of forests following the use of forests for carbon sequestration as a measure to mitigate climate change (see ► Chap. 6). The economic value of a forest is therefore not only based on trees as a source of timber or biodiversity, but also as a potential site for carbon storage. Countries with huge forests, such as the Democratic Republic of Congo, Tanzania and Brazil, have therefore received payments via international programmes to reduce climate change through forest protection. However, information about the increased value of forests and the actual payments will not necessarily be available to all people affected by these programmes. This may be due to practical difficulties of disseminating the information, but government officials in the responsible ministries may also choose to hold back on information or selectively inform forest communities. Thus, people who live in forest areas do not necessarily receive information about how much their area is worth, and what they need to do to take part in the income from protecting the forest. Some of these funds may therefore benefit government officials and non-governmental organizations more than forest communities.
7. The use of violence and coercion is a means of power that can be applied by the state more or less in agreement with national legislation. An example is the forced relocation of people in connection with mining or conservation. In many countries in the Global South, there have in recent years been an increased militarization of conservation for instance in and around national parks (► Chap. 4). Many incidences have been documented of people who have been beaten up or even shot. In addition, violence is employed by other actors that operate outside the law, such as criminal networks or resistance movements including 'jihadist' groups (► Chap. 8).
8. 'The Weapons of the Weak' is a form of power resource described by Scott (1985). In examples of asymmetric power, Scott shows how people have their methods to resist exploitation, and that such resistance can actually have a substantial impact. This implies that people may pretend to support a project,

but in practice, they resist it through various passive forms of everyday resistance (► Chap. 2).

9. Finally, power resources presented above are often unequally distributed within categories such as ethnicity gender, age, class and nationality. This implies, for instance, that some ethnic groups tend to have small chances to win struggles over land and natural resources (this constitutes a basic assumption behind Avatar), and in ► Chap. 5 we provide examples of how women have been excluded in decisions about environmental governance.

In political ecology, all nine types of power resources, as well as others, are discussed in studies of environmental conflicts.

► The Power Struggle on Pandora

Avatar can be seen as a story about a power struggle over environmental governance. The intention of the mining company is to transport the mineral unobtanium from Pandora to Earth. In order to achieve this aim, the company has used its economic power resources to establish both a military unit and a research unit.

The company has power resources at its disposal in the conflict with the Na'vi that make it seem completely invincible. It is difficult to imagine that the company would adjust itself to the Na'vi's local knowledge, environmental management and decision-making. Instead, the intruders try to use the knowledge they can gain about the Na'vi as well as military power resources to forcefully remove the local inhabitants. Seen from the Na'vi, the war takes place in their own environment, and they possess crucial knowledge about nature. Furthermore, marine soldier Jake Sully's knowledge of the enemy's fighting methods appears to be useful.

In order for Jake to lead the fight for the Na'vi, he must acquire authority and political power on their terms. In Na'vi history, there have been five great leaders who have tamed and flown a great dragon bird called *Toruk*. When Jake manages to repeat this achievement, he is appointed their leader. The breakthrough in the fight comes when the nature goddess Eywa helps by mobilizing Pandora's wildlife.

The parties obviously fight about two very different forms of environmental governance on Pandora, and they use indeed very different power resources to try to achieve this. ◄

1.7 Seventh Synthesis: Linkages between Different Geographical Levels

A central scope in political ecology is to try to understand how everyday life and local environmental conflicts are not only influenced by local conditions, but that structural factors and actors at both national and global levels also can have great influence at the local level. In their seminal contribution to political ecology, Blaikie and Brookfield (1987) show how power relations in environmental change can be studied starting with the agency of a local land manager (for example a farmer) through various

1 *chains of explanation* via the social unit the farmer is part of (family, village, etc.), through the local administration, central state administration, and the national and global economy (see ▶ Chaps. 2 and 5). Later contributions to political ecology have pointed out that there are not always linear chains of explanations and that influences may often better be charted within *networks* of various actors (Robbins 2004; Rocheleau 2008).

Linkages between different levels stand in contrast to a focus only on local factors, and where influences from external actors and structures are ignored. On the other hand, linking the levels provides a richer perspective than paying attention only to external factors.

From Pandora to the Earth

In *Avatar* a situation is presented where the local communities meet a threat that obviously comes from the outside. However, the film does not provide insight into the conditions on the Earth that have resulted in the colonization of Pandora. Only once Jake Sully briefly mentions that the natural environment on Earth has been destroyed. In political ecology studies of real communities, that are threatened by the activities of external actors such as mining companies, researchers would look for causes in the political economy that led to new processes of resource appropriation. Furthermore, scholars would also examine local and national factors that could have contributed to the situation.

1.8 Eighth Synthesis: Temporal Connections

Time constitutes an important aspect in political ecology. This implies both an interest in specific environmental questions as they are played out *today*, as well as historical analyses of how and why a particular situation has occurred. Two main types of historical studies play an important role here. First, *environmental data* are often collected, where researchers attempt to cover as long a time period as possible, to understand the background to ongoing trends. The longer time series, the better. Long time series make it possible to describe historical changes in landscapes and natural resources. Secondly, in studies of *conflicts*, it is also important to have an historical approach to understand how and why the conflicts have evolved.

From the Earth to Pandora

If *Avatar* had been a story from real life, and James Cameron were a political ecologist, the movie would have given us insights into some important developmental characteristics about the Earth that would provide us with the historical background to the colonization of Pandora.

1.9 Ninth Synthesis: Linking Different Types of Knowledge and Scientific Methods

In the previous sections, we have presented political ecology as an approach where researchers seek to connect elements of different forms of knowledge, to best possibly understand how and why people manage nature in specific ways. This receptiveness to various knowledge elements implies the necessity also of involving those research methods that in each case seem to be best suited for collecting and analysing different forms of knowledge. We have already introduced the use of natural science methods during the first synthesis. In the following, we will therefor only look at social science methods.

While a great deal of environmental research usually applies quantitative methods, qualitative methods play the leading role in political ecology. In order to be able to describe, explain and make normative assessments of various types of environmental governance and related conflicts, it is almost always necessary to apply qualitative methods. This is partly because the understanding of the total picture has to be pieced together by different bits of knowledge collected from interviews with different actors and written sources. It would be meaningless to distribute questionnaires with identical questions if the purpose is to learn about different forms of participation in networks of different actors. It is better, for example, to conduct semi-structured interviews with questions that are adapted to each of the interviewees. Moreover, political ecologists are often concerned about acquiring a good understanding of ways of thinking through the use of ethnographic methods in studying environmental governance rather than quantifications based on questionnaires with pre-defined answer alternatives. Nevertheless, quantitative methods in political ecology research can often provide important knowledge in combination with qualitative data.

The Mixed Methods at Pandora

As mentioned in the first synthesis, Jake Sully's research unit studies everything that has to do with the Na'vi, in addition to all kinds of aspects of nature on Pandora. The social science part, however, seems to concentrate entirely on participatory observation and do not involve other qualitative or quantitative methods. In order to get acceptance to stay around the Na'vi, the researchers go a great deal further than any real-world researcher would be able to. The Pandora researchers have developed Na'vi bodies (avatars) that they enter into and thereby look just like the Na'vi.

1.10 Tenth Synthesis: Critical and Constructive Contributions

The final synthesis is about what political ecologists do with their findings. Political ecology is a critical approach. This implies a critical view on power as well as questioning many aspects that tend to be taken for granted. Numerous political ecology studies have focused on loss of environmental sustainability, injustice and violation of human rights. At the same time, political ecologists also provide constructive suggestions about *which* changes should be made to create a more just and sustainable world. Paul Robbins calls the critical and the constructive side of political ecology for *hatchet* and *seed* (Robbins 2012). Peter Walker, has argued that political ecologies should put more emphasis on the seed dimension (Walker 2006), and Piers Blaikie (2008) agrees and criticizes other political ecologists for stating that they do not feel any responsibility for being *useful.*

Hatchet and Seed on Pandora

In *Avatar*, Jake Sully and his research unit takes different positions in three different phases. In the first phase, they contribute neither critical nor constructive knowledge to the Na'vi. Instead, they provide knowledge that the mining company and its military force need for their intentions to outwit the Na'vi. In his daily video log Jake Sully says that there is nothing they have that the Na'vi want, and that he feels that everything that he has been told to do is a waste of time. The Colonel follows this log, and understanding that Jake is about to change side he decides to immediately implement a military offensive against the Na'vi clans and destroy their enormous 'Home tree'.

The second stage is a short in-between phase that occurs when Jake and his research colleagues realize that the military is preparing to attack. The researchers now provide a critical contribution, in which they attempt to convince the mining colony's military and economic leadership to call off the attack because it would be a threat against the Na'vi as well as the natural environment on Pandora. But the researchers fail to get any approval for this from the colonel and the director of the mining company.

In the third phase, there is full war, and the researchers participate on the Na'vi's side. Here the researchers obviously go much farther than providing critical and constructive perspectives. We think it is great to watch a movie where researchers are for once not portrayed as comic nerds or as thoroughly cynical and evil. In *Avatar,* they instead take on the role as good helpers to a people who are threatened by ecocide.

For political ecologists there is a large scope of action between the two extreme points of taking part in armed conflicts or not to speak out after having uncovered a situation of for instance injustice. Instead of supporting any of these extremes, we believe that political ecologists, like all scholars, have an ethical duty to speak out when our research reveals oppression, injustice or destruction of nature. This is the critical side of political ecology, and it can be combined with constructive suggestions for alternative politics.

One problem is that research does not necessarily lead to improved conditions. For example, political ecologists have often critiqued environmental governance without any noticeable political impact, because the critique does not fit the agenda of the bureaucrats or politicians in charge. Knowledge production that raises fundamental questions about the premises behind policies is often considered irrelevant and thus excluded from having any political impact. Decision makers instead demand instrumental knowledge that they can immediately use, and that does not threaten their own position. Therefore, there may be limited opportunities for critical research to be used directly (Nustad and Sending 2000). However, in a longer time perspective critical research may contribute to changing perspectives and political agendas.

Another problem is that critique may also involve a personal risk for the individual researcher in terms of loss of further research funding or job opportunities.

Political ecology has increasingly gained more ground as a university subject during the last 20 years, especially in the English-speaking world, but it still has a limited political influence. 'Apolitical approaches' continue to dominate in providing premises for policy-making (► Chap. 2).

Methods: What Research Questions to Ask?

Research questions are usually key tools in research projects, including student papers and dissertations. After deciding on a topic, and perhaps after also selecting a more narrow focus on a case, students need to think about research questions.

We can divide research questions into several categories, although there may also be combinations between these. We will here highlight three types of research questions that are often asked in political ecology studies; descriptive, explanatory and normative.

Most methodology books seem to agree that descriptive and explanatory questions are the most important ones, and some even argue that these are the only legitimate research questions (e.g. White 2017).

Descriptive questions aim at identifying what aspect to focus on in characterizing a topic or a case. When answering descriptive questions, knowledge is gathered as basis for further investigations. Descriptive questions may start with 'what', 'who', 'where', 'when' and 'how'. Students are often expected to formulate at least one descriptive question in their papers or dissertations. At the same time, they are usually also expected to ask at least one other type of research question building on insights gained from descriptive questions.

If a student chooses to focus on the case of mining, a descriptive question could, for instance, be about whether the mine has positive or negative impacts for local inhabitants or for mine workers. And this could again lead to specifications of subordinate questions about different types of impacts.

Explanatory questions may start with 'why' or 'how' and be articulated to focus on causes explaining the situation discovered through descriptive questions. This could, for instance be a question about the causes behind the establishment of the particular mine, and which actors took part in the process of deciding on the investment and with the use of which power resources. There could also be a question to explain, for instance, how a mining company was able to appropriate land from farmers. Another question could be why the mining activity resulted in particular environmental problems such as polluting the ground water. Were these unavoidable impacts of this type of mine, or were they also results of weak environmental legislation or policy-making?

Normative questions are held to be acceptable by some methodology text books, for instance Alvesson and Sandberg (2013). In contrast, White (2017) argues that normative questions should be avoided as research questions, because there is no single correct answer to them, and they require the researcher to make an ethical judgement.

However, the Frankfurt School (see ▶ Chap. 2) and other traditions associated with critique of positivism reject the claim that social science research should be value free. Instead, scholars should explicitly discuss values, including their own. This means that transparency rather than objectivity is an epistemological goal. Political ecology is a critical approach, and normative questions play a key role in this research.

We can distinguish between two types of normative questions. First, they may be asked to assess a situation compared to specific standards, such as targets for environmental justice, sustainability or human rights. This implies that scholars or students may discuss similarities and differences between their own empirical knowledge or available data compared to the chosen norms or principles.

Secondly, normative questions can be asked to investigate what can be done to improve a problematic situation. Students may, for instance, start with asking a descriptive question about what solutions have been proposed to stop pollution from mining, and proceed to discuss strengths and weaknesses of suggested solutions. Such studies could be examined through analyses of documents and media texts about local impacts of a particular mine or about a mining activity in a particular country (e.g. gold mining in Ghana, uranium mining in the USA, or copper mining in Chile), or even at a global level. Such normative questions could also be studied by collecting data through methods such as interviews, observations or written texts, and by examining data from natural science studies.

It is useful to start the work of a paper or dissertation by formulating tentative research questions. This will provide some direction for where to go. Nevertheless, along the process, it may be important to modify or change the original research questions in order to adjust them better to the available empirical material as well as to theorical frameworks that can be used in the analysis. When elaborating the outline for a thesis, it is useful to specify elements to address each research question, and decide what methods to use, and, for instance, what particular questions to outline in interview guides for different groups of interviewees.

See also the methodology box in ▶ Chap. 3 on how research questions may be formulated about narratives and discourses.

1.11 Delimitations of this Book

We argue that political ecology, as well as social science in general, benefit when researchers possess a broad range of perspectives, theories and concepts as possible building blocks they can draw on when designing empirical studies. In this chapter we present our selection of some main building blocks of contemporary political ecology.

Is it really possible, in specific studies, to apply perspectives from all the ten syntheses outlined above? There are certainly few people who get the chance to conduct extensive studies where they can really go deeply into all these aspects. It will always be necessary to make practical delimitations of scope. However, for students and scholars, political ecology offers a starting point with a rich and nuanced combination of perspectives, and from these students can choose to emphasize specific aspects in their own research.

When we have listed these ten syntheses, we are also aware that there are other themes, theories and approaches that are applied in political ecology that also could have been highlighted in this first book chapter. This book therefore gives an introduction to *our* approach to political ecology, and how we as scholars understand and use this approach. The examples and case studies we present throughout the book are therefore also mostly taken from our own research. There are therefore themes within and perspectives on political ecology that we do not deal with or only briefly mention, such as urban political ecology and Science and Technology Studies.

1.12 The Other Chapters in this Book

As mentioned earlier, ▶ Chap. 2 is a theory chapter where we go through the main aspects of the short history of political ecology. ▶ Chapter 3 presents *bioprospecting*, which is an economic activity where companies (usually in the Global North) elaborate new market products from biological samples and sometimes also based on traditional knowledge about use (often in the Global South). In ▶ Chap. 3 we also provide a conceptual framework for studying discourses and narratives, and we compare claims within these social constructions with research knowledge about practices. Our discourse-practice perspective is also demonstrated in several of the subsequent chapters.

In ▶ Chap. 4, we discuss the establishment and social impacts of protected areas and show examples of a gap between a widespread discourse of community-based conservation and practices of centralized and top-down governance. ▶ Chapter 5 introduces feminist political ecology. We present examples of the application of this approach in studies of the establishment and management of protected areas.

▶ Chapter 6 is about approaches to mitigate climate change, with an emphasis on efforts to Reduce Emissions from Deforestation and forest Degradation in developing countries (REDD). ▶ Chapter 7 is about two classic themes in political ecology – degradation and marginalization. We show how pastoralists are marginalized politically and economically through the use of different arguments that constitute part of a modernization discourse. In addition, we point out how claims about environmental degradation with poor empirical foundations and blamed on pastoralists have contributed to this marginalization.

In ► Chap. 8, we look critically at the idea that scarcity of natural resources and climate change will result in more violent conflicts. ► Chapter 9 deals with agricultural development and particularly how population growth and markets affect possibilities for sustainable agriculture in Africa. This chapter also includes a discussion of ongoing processes of 'land grabbing', where investors obtain control over large land areas in the Global South. In ► Chap. 10, we summarize experiences and insights from the type of political ecology that we have used in the case studies, and we suggest some possible ways forward for a political ecology for the future.

We have in this first chapter used the environmental conflict on the fictional planet Pandora as an example to illustrate the content of the ten syntheses that we consider key in political ecology. The rest of the book is about real life on the real planet Earth. We use examples in the following chapters that we know from our own research about environmental governance - mainly from Africa and Norway, but also from elsewhere. This implies that the book points out several similarities and differences between environmental governance in the Global South and North and how they are inter-connected. In addition, we are concerned with presenting the various forms of global dimensions that characterize environmental governance in different locations. Some examples are also drawn from other parts of the world, and the topics and approaches that we present in this book are relevant for planet Earth in general.

Finally, some remarks on the formats we use in this book. Most chapters start with a trailer introducing the chapter topic often with a short story from the field or from popular culture before it continues with an overview of the main questions discussed in the chapter. Chapters also contain Case Studies, which give empirical illustration mostly from our own research of the questions or theoretical concepts presented in the chapters. There are also a small number of Examples, which are much shorter than the Case Studies. In some chapters, we have also included Boxes with discussions of methods or methodological approaches that can be used in political ecology, and we give Definitions of some key concepts. Finally, in all chapters, except this chapter and the final one, we end with some Conclusions that sum up the messages in the chapter before we list some key Questions that students can discuss in class or answer individually.

? Questions

1. While the movie Avatar has been embraced by indigenous activists, it has also been criticized for 'white saviourism'. What is your own view? (It is useful to watch the movie first.)
2. Imagine that you are a researcher starting up a study on Pandora:
   a. Formulate research questions based on the types presented on pages 23–24.
   b. Select three of the syntheses in this chapter and discuss how the research project could draw on each on them.

References

Adams, W., and M. Mulligan, eds. 2003. *Decolonizing nature: Strategies for conservation in a post-colonial era*. London: Earthscan.
Agrawal, A. 2005. *Environmentality: Technologies of government and the making of subjects*. Durham: Duke University Press.
Alvesson, M., and J. Sandberg. 2013. *Constructing research questions*. London: SAGE.
Amin, S. 1977. *Imperialism and unequal development*. Hassocks: Harvester Press.
Batterbury, S., T. Forsyth, and K. Thomson. 1997. Environmental transformations in developing countries: Hybrid research and democratic policy. *Geographical Journal* 163 (2): 126–132.
Blaikie, P. 2008. Epilogue: Towards a future for political ecology that works. *Geoforum* 39 (2): 765–772.
Blaikie, P., and H. Brookfield, eds. 1987. *Land degradation and society*. London and New York: Methuen.
Dahl, R. 1957. The concept of power. *Behavioral Science* 2: 2001–2015.
Engelstad, F. 1999. *Om makt: teori og kritikk*. Oslo: Gyldendal Norsk Forlag.
Fletcher, R. 2010. Neoliberal environmentality: Towards a poststructuralist political ecology of the conservation debate. *Conservation and Society* 8 (3): 171–181.
Flyvbjerg, B. 1998. *Rationality and power: Democracy in practice*. Chicago: University of Chicago Press.
Forsyth, T. 2003. *Critical political ecology. The politics of environmental science*. London: Routledge.
Foucault, M. 2008. The Birth of Biopolitics: Lectures at the Collège de France 1978–1979. New York: Picador.
Frank, A.G. 1979. *Dependent accumulation and underdevelopment*. New York: Monthly Review Press.
Giddens, A. 1984. *The constitution of society, Outline of the theory of structuration*. Cambridge: Polity Press.
Goldman, M.J., P. Nadasdy, and M.D. Turner, eds. 2011. *Knowing nature: Conversations at the intersection of political ecology and science studies*. Chicago: University of Chicago Press.
Gramsci, A. 1932/1975. *Letters from prison*. New York: Harper Colophon.
Long, N. 1992. From paradigm lost to paradigm regained? The case for an actor-oriented sociology of development. In *Battlefields of knowledge. The interlocking of theory and practice in social research and development*, ed. N. Long and A. Long. London and New York: Routledge.
———. 2001. *Development sociology: Actor perspectives*. London: Routledge.
Lukes, S. 2005. *Power: A radical view*. 2nd ed. New York: Palgrave Macmillan.
Martin, A., B. Coolsaet, E. Corbera, N.M. Dawson, J.A. Fraser, I. Lehmann, and I. Rodriguez. 2016. Justice and conservation: The need to incorporate recognition. *Biological Conservation* 197: 254–261.
Neumann, R.P. 2005. *Making political ecology*. London: Hodder Education.
Newitz, Annalee. 2009. When will white people stop making movies like 'Avatar'? 18 Dec 2009. http://io9.com/5422666/when-will-white-people-stop-making-movies-like-avatar. Visited 29 June 2020.
Nustad, K, and O.J. Sending. 2000. The instrumentalization of development knowledge. In: Stone, D. (ed). Banking on knowledge. London: Routledge.
Paulson, S. and L. Gezon (eds). 2005. Political Ecology Across Spaces, Scales and Social Groups. New Brunswick, NJ: Rutgers Universtiy Press.
Pretty, J.N. 1995. Participatory learning for sustainable agriculture. *World Development* 23 (8): 1247–1263.
Robbins, P. 2004. *Political ecology. A critical introduction*. Oxford: Blackwell.
———. 2007. *Lawn people: How grasses, weeds, and chemicals make us who we are*. Philadelphia: Temple University Press.
———. 2012. *Political ecology: A critical introduction*. 2nd ed. Oxford: John Wiley & Sons.
———. 2020a. *Political ecology. A critical introduction*. 3rd ed. Oxford: Wiley Blackwell.
———. 2020b. Is less more … or is more less? Scaling the political ecologies of the future. *Political Geography* 76: 102018.

Rocheleau, D. 2008. Political ecology in the key of policy: From chains of explanation to webs of relation. *Geoforum* 39 (2): 716–727.

Russel, B. 1938. *Power: A new social analysis*. London: Allen & Unwin.

Scott, J.C. 1985. *The weapons of the weak: Everyday forms of peasant resistance*. New Haven: Yale University Press.

Stones, R. 2005. *Structuration theory*. Houndmills: Palgrave Macmillan.

Survival International. 2010. 'Avatar is real', say tribal people. http://www.survivalinternational.org/news/5466. Visited 29 June 2020.

Svarstad, H., L.K. Petersen, D. Rothman, H. Siepel, and F. Wätzold. 2008. Discursive biases of the environmental research framework DPSIR. *Land Use Policy* 25 (1): 116–125.

Svarstad, H., T.A. Benjaminsen, and R. Overå. 2018. Power theories in political ecology. *Journal of Political Ecology* 25: 350–363.

United Nations. 2015. *World fertility patterns 2015. Data booklet*. New York: United Nations.

van der Horst, H. 2010. Avatar and the Racism of Virtue: The Na'vi are Americans (in a reverse mirror image). https://www.academia.edu/1549435/Avatar_and_the_Racism_of_Virtue.pdf. Visited 29 June 2020.

Walker, P.A. 2006. Political ecology: Where is the policy? *Progress in Human Geography* 30 (3): 382–395.

Weber, M. 1964 (1947). *The theory of social and economic organization*. New York: Free Press.

White, P. 2017. *Developing research questions*. 2nd ed. London: Palgrave Macmillan.

World Commission on Environment and Development (WCED). 1987. *Our Common Future*. Oxford: Oxford University Press.

Theoretical Influences and Recent Directions

Contents

T. A. Benjaminsen, H. Svarstad, *Political Ecology*, https://doi.org/10.1007/978-3-030-56036-2_2

2

Overview

Although political ecology (PE) has a relatively short history in research and university teaching, the field is the result of a range of theoretical influences. In this chapter, we introduce some of the main schools or ways of thinking that have influenced political ecology since the inception of the field in the 1970s; Marxist political economy, human and cultural ecology, poststructuralism, peasant studies, and critical theory.

The chapter also introduces several key concepts and theories that current political ecology draws on as well as directions within the field and related approaches such as environmental justice and degrowth.

2.1 Marxist Political Economy

Political economy is a term that is used for various scholarly approaches at the interface of politics and economics. In political ecology, the political economy tradition established by Karl Marx (1818–1883) has played a key role. Marx was concerned with the organization of economic production within the capitalist system that he saw emerging in Europe. He compared this with pre-capitalist and other modes of production and with communism that he thought would replace capitalism. It is not the political project of Marx or his anticipation about future development that is important here, but rather the conceptual tools for social analysis he created to understand the workings of Western capitalism.

Definition

According to a Marxist conceptual framework, a *mode of production* consists of relations of production and productive forces. Relations of production primarily include property rights and other power resources that enable decisions about production. Productive forces are composed of natural resources, technology and labour. Capitalists own the means of production, sell products on the market and accumulate capital through appropriating the profits, which are used for new investments in addition to increased wealth and consumption, while the proletariat is forced to sell their labour. Marx was particularly concerned about how capitalism is based on an exploitation of cheap labour, and he also briefly mentions the exploitation of nature.

Some Marxists have since imagined that such forms of exploitation occur especially in the meeting point (the so-called articulation) between capitalism and other modes of production (e.g. 'the African') (Coquery-Vidrovitch 1975; Meillassoux 1981; Foster 1999). This has been called the Articulation of Modes of Production Approach, which is a Marxist branch of dependency theory (see the fourth synthesis in ► Chap. 1).

Class relations under capitalism and the persisting powers reproduced by these relations represent an important focus in Marxist theory (Isaac 1987). Human agency is at the centre of Marx' conception of power, but it is an agency that is socially conditioned, which is reflected in this famous quote:

> Men make their own history, but they do not make it as they please; they do not make it under self-selected circumstances, but under circumstances existing already, given and transmitted from the past (Marx 1852: 5).

Hence, a Marxist power theory understands human agency as constrained and to a large extent produced by historically established structures in a society. While structure generates the potential and limits for the exertion of power, agency tends to reproduce structure. To illustrate this point, Isaac (1987: 81) uses the example of a capitalist such as David Rockefeller (1915–2017), who certainly was a powerful man.

> But a social theory of power must explain what kinds of social relations exist and how power is distributed by these relations, such that it is possible for David Rockefeller to have the power that he has. To do this is not to deny that it is *he* who possesses this power, nor to deny those personal attributes determining the particular manner in which he exercises it. It is simply to insist that the power individuals possess has social conditions of existence, and that it is these conditions that should be the primary focus of theoretical analysis (Isaac 1987: 81).

Michael Watts (1983) provides an early example of such a power analysis in political ecology by focusing on how historically produced social structures condition the agency of individual smallholders in his study of small-scale farming in northern Nigeria. He concludes that commodification caused starvation and economic marginalization among peasants who became increasingly dependent on an unstable market. This market integration led them to become more vulnerable, and they therefore had to take up loans and generally take more risks. Previously self-sufficient, peasants gradually became underpaid farm workers. This in turn led to decreasing investments of labour on their own land, resulting in the degradation of soils on land where food crops were grown. Hence, this was a largely structural explanation of processes of deprivation and soil degradation.

Definition

In addition to Watts' prominent contributions to political ecology, David Harvey's Marx-inspired work has been influential in the social sciences in general, including in political ecology, during the last two or three decades. One example is Harvey's modification of Marx's notion of '*primitive accumulation*', which Marx saw as a key feature of how capitalism works. This refers to a historical process of divorcing the producer from the means of production through privatization of commons. According to Harvey (2003: 149),

> primitive accumulation as Marx described it ... entailed taking land, say, enclosing it, and expelling a resident population to create a landless proletariat, and then releasing the land into the privatized mainstream of capital accumulation.

In Chap. 26 of volume 1 of *Capital*, Marx famously used the enclosures in England in the eighteenth and nineteenth centuries as an example of primitive accumulation (Marx 1990). Many smallholders who lost access to land in this process became labourers in the emerging textile industry in cities like Manchester and Liverpool. The enclosures in England took place mostly between 1750 and 1850 when about 4000 individual enclosures were enacted in Parliament (Mingay 1997).

Primitive accumulation may, however, easily be understood as a historical process that belongs to a distant past. This is why David Harvey has introduced the term *accumulation by dispossession* to point out that the combination of dispossession of vulnerable people and capital accumulation is currently an ongoing process around the world. Hence, there is nothing 'primitive' about this process. In political ecology, the introduction of this term has sparked a renewed interest in the combination of dispossession and capital accumulation (Büscher 2009; Li 2009; Corson 2011; Kelly 2011; Benjaminsen and Bryceson 2012).

Accumulation by dispossession is, however, not a new phenomenon in the Global South. It started during colonialism. But its pace has probably increased with economic globalization. In several chapters in this book we give examples of this process: biodiversity conservation (▶ Chap. 4), forest conservation as climate mitigation (REDD) (▶ Chap. 6), and large-scale agricultural investments (▶ Chap. 9). States, business interests or big conservation organizations accumulate capital, while smallholders tend to be dispossessed of access to land and natural resources. Landscapes and 'wilderness' are valued, while the people who have historically inhabited these areas are seen to be in the way.

Consequently, non-capitalist spaces and resources are opened up for accumulation. Tania Murray Li has developed this idea further by proposing that it is the places and the resources in these places that are valued, 'but the people are not, so that dispossession is detached from any prospect of labour absorption' (Li 2009: 69). These places might be interesting for investments in large-scale production of food crops or biofuels—or for the development of 'ecotourism' such as safari tourism. But local people are in the way of such investments. Their labour is not needed in contradiction to the classic example of the double process of enclosures and industrialization in England.

Definition

Harvey (1999) has also introduced the concept *spatial fix* that has increasingly been used as an analytical tool in political ecology. This refers to the tendency of a falling rate of profit in capital investments, which again leads to a crisis with over-accumulation of capital (in the form of reduced investment opportunities in the geographical space in question) as well as reduced need for labour (leading to increased unemployment). The solution for capitalism is a spatial fix, which implies moving surplus capital into new spaces and new markets. Rather than being permanent, these fixes should be viewed as temporary remedies solving a crisis of accumulation. As Harvey (2014: 7) notes, instead of solving its underlying contradictions, capital 'has the nasty habit of simply moving them around', with the state playing a key supportive role.

The lens of spatial fix has been used to discuss the role of scale in political ecology (Brown and Purcell 2005; Cohen and Bakker 2014) and, for instance, to explain how the movement of capital to invest in African agriculture produces winners and losers (Bergius et al. 2018), and to explain the growth in aquaculture as a fix for industrial overfishing (Mansfield 2010).

More recently, political ecologists have extended this idea further to not only include the production of space, but also the production of nature within 'socio-ecological fixes' (Ekers and Prudham 2015). This implies studying more closely how landscapes and ecologies are transformed as a result of capital moving into new spaces as well as creating new commodities such as carbon, biodiversity, ecosystem services and genes. Some of the political ecology literature studying socio-ecological fixes also connects these fixes directly to processes of accumulation by dispossession.

Definition

Commodity fetishism is another concept from Marx that has been frequently used in political ecology analyses. A fetish is an object that is given religious value, and the origin of this value is therefore of a mystical character. Commodity fetishism was used by Marx to highlight that there is often an ideological veil obscuring our understanding of capitalistic production. This means that commodities tend to appear as objects detached from the social relations they are produced under. The concept 'commodity fetishism' is meant to highlight the fact that there in a capitalistic economy tends to be an information gap between production and consumption. This gap widens the more global the commodity becomes. The consumer therefore knows little about the social and environmental conditions of production of particular commodities. In a global economy based on mass production, it is for instance difficult for consumers to know whether the commodities they buy have been produced according to certain ethical and environmental practices.

Therefore, a consumer cannot know whether clothes from a specific manufacturer are produced using child labour before an organization for labour or human rights investigates this and the media writes about it. Similarly, it is difficult to know whether the palm oil in the peanut butter comes from sustainable small-scale producers in West Africa or from plantations in Indonesia that have destroyed habitats for orangutans. Even if some of us have the opportunity to go on a safari in Africa, it is hard to know if the hotel where we stay really has such a harmonious relationship with local communities as the hotel management claims, or whether people have been dispossessed and lost livelihood sources as a result of tourism.

As we will show in ► Chap. 6, carbon is a new commodity where there is such a wide information gap between production (e.g. tree planting or forest conservation) and consumption (purchase of climate credits). Commodity fetishism is therefore a concept that can be useful to use for instance in analyses of climate mitigation projects involving the carbon market.

Hall et al. (2011) represent another clearly Marx-inspired analysis within political ecology. They focus on the changing ways that people are excluded from land in Southeast Asia. Processes of modernization associated with economic growth, industrialization and urbanization have generally led to de-agrarianisation, which means that 'agriculture becomes progressively less central to national economies and to the livelihoods of people even in rural areas' (Hall et al. 2011: 1). In this context of de-agrarianisation, exclusion in various forms takes place. But '[e]xclusion is not a random process, nor does it occur on a level playing field. It is structured by power relations' (Hall et al. 2011: 4). There are a number of 'powers of exclusion' at play. Hall et al. (2011) label these as 'licensed exclusions' (land titling and reform), 'ambient exclusions' (driven by environmental conservation), 'volatile exclusions' (driven by booms in some cash-crops), 'post-agrarian exclusions' (urban expansion and other conversions of agricultural land to non-agricultural use), 'intimate exclusions' (smallholders' exclusions of neighbours and kin), and 'counter-exclusions' (resistance and reactions from below).

Finally, Marxist political economy has also, since the 1970s, provided a firm critique of the influential ideas of Thomas Robert Malthus (1766–1834) on the role of population growth. According to Malthus (1798), increasing population pressure on natural resources is the paramount cause of human misery. His 'population law' postulated that the population growth of the poor inevitably will exceed the resource base represented by food production and cause widespread hunger and poverty. Marx later strongly rejected this idea, which he saw as an attempt to defend class relations under capitalism and blame poverty on the poor themselves (Ross 1998).

Since the 1960s, the ideas of Malthus have seen a revival and are frequently used by scholars, policy-makers and the media to explain environmental problems. Neo-Malthusian ecologists such as Paul Ehrlich and Garrett Hardin were among the first leading scientists to use Malthus on the environment. Ehrlich's book *The Population Bomb* (Ehrlich 1968) and Hardin's article *The Tragedy of the Commons* (Hardin 1968) have in particular been influential. Early contributions to political ecology consisted essentially of Marxist critiques of such views of population growth as the main factor behind environmental decline.

'Poor people make poor land' was, however, a powerful slogan accepted by both Neo-Malthusian and Marxist inspired scholars alike. This slogan was also repeated by the Brundtland report (WCED 1987), which used it as a political argument for the need to alleviate poverty in order to arrest environmental degradation. Hence, poverty was framed as the main cause of environmental decline and pollution.

Early Marxist critiques of Malthusianism reflected in some way Marx's idea that '… all progress in capitalist agriculture is a progress in the art, not only of robbing the worker, but of robbing the soil' (Marx 1990: 638). Hence, market integration and the expansion of capitalism would automatically lead to environmental degradation, a link that was taken for granted and rarely investigated empirically in early Marxist contributions to political ecology.

However, while Marxist critics tended to accept the environmental impacts of human production described by the various Neo-Malthusian reports, they pointed at the inherent lack of social and political analysis in these studies and argued that Malthusian thinking would tend to lead to policies of 'blaming the victims'. In this way, as pointed out by Enzensberger (1974), ecologists and other natural scientists may pretend to be 'objective' and 'apolitical', but they become political actors when engaging in environmental debates, which are essentially about political choices with winners and losers. The presumed neutrality of ecology as a science is therefore illusory, and what Enzensberger referred to as 'political ecology' corresponds to what political ecologists later have called 'apolitical ecology' (Robbins 2020a).

Political ecologists have, however, since the 1990s challenged many environmental degradation narratives. While Marx' idea about 'the robbing of soils' in capitalist agriculture should be critically investigated like any 'taken-for-grantedness', and while there is not an obvious connection between environmental degradation and political-economic marginalization, there has recently been a revival of Marxist thinking in political ecology. Instead of marginalization necessarily leading to degradation, political ecology studies have documented how degradation narratives serve to justify elite capture and smallholders' dispossession of land and natural resources including forms of accumulation by dispossession (Benjaminsen 2015; Bergius et al. 2020; Cavanagh et al. 2020). In this way, degradation narratives may contribute to driving land-use conflicts rather than population growth and environmental degradation as in the Malthusian scenario.

2.2 Human Ecology and Cultural Ecology

Human Ecology and Cultural Ecology are closely related approaches within interdisciplinary human-environment and nature-society studies although with somewhat different origins (Zimmerer 2013).

Human Ecology was coined by geographers at the University of Chicago in 1907 under the influence of well-known ecologists such as Frederic Clements, but has later been mostly championed by social scientists. *Cultural Ecology* also emerged in the early twentieth century in the USA, but as a theory of cultural change focusing on human adaptation to material conditions (economic and environmental) (Zimmerer 2013).

In both approaches, human society, often at a local community scale, are studied as part of an ecosystem with a focus for instance on flows of energy, nutrients or calories. Research on risk, systems analysis, resilience and adaptation are important directions within this tradition.

A textbook in Human Ecology (that embraces both Human and Cultural Ecology) states that this approach 'emphasizes the complex ways in which humans shape and are shaped by their environment' (Bates and Tucker 2010: 1). While earlier human ecology 'tended to ask how traditional behaviors enabled a population to maintain itself in a specific environment, … today we (human ecologists) are more likely to ask: What are the problems confronted by the populations at this place? How do individual actors deal with them?' (Bates and Tucker 2010: 4).

In such analyses, key concepts from the discipline of ecology are used such as resilience, stability, equilibrium, non-equilibrium, carrying capacity, habitat, niche, territory, food chains and energy flows. But social science themes such as division of labour, social hierarchies and property rights are often also discussed.

In particular, cultural ecology's focus on smallholders has been influential in political ecology. Cultural ecologists, such as Robert Netting (1934–1995), have demonstrated how small-scale producers operate within complex and variable systems, but manage to navigate these systems in rational and comprehensible ways (Robbins 2020a). Netting carried out detailed empirical studies of smallholder agriculture in Switzerland and Nigeria. These studies resulted in three classics in cultural ecology; *Cultural Ecology* (1977), *Balancing on an Alp* (1981), and *Smallholders, Householders* (1993).

Netting's main thesis was that smallholders possess sufficient knowledge, motivation and ecological insight to manage land and environments in sustainable ways. He also demonstrated that small-scale agriculture can be intensified and efficient. While small-scale producers are often seen from the outside as primitive, conservative and inefficient, Netting argued that it is rather the modern state that is primitive, especially in the way it treats smallholders.

Definition

The methodological approach called *progressive contextualization* represents another example where cultural ecology has influenced political ecology. This approach was introduced by Andrew Vayda (1983), who is another prominent human and cultural ecologist, for instance as the founder of the journal *Human Ecology*.

Vayda illustrates progressive contextualization by referring to a study where

> a goal of the investigators was to understand the forces contributing to deforestation in the Indonesian province of East Kalimantan on the island of Borneo. To achieve this goal, they started by focusing on specific activities, such as timber cutting, performed by specific people in specific places at specific times. They then traced the causes and effects of these activities outwards. In doing so, they remained committed to the holistic premise that adequate understanding of problems can be gained only if they are seen as part of a complex of interacting causes and effects. But the investigators avoided *a priori* definitions of the boundaries of such complexes - for example, that they correspond to the boundaries of an ecosystem or of a human community. (Vayda 1983: 266)

Clearly inspired by progressive contextualization, political ecologist Piers Blaikie and cultural ecologist Harold Brookfield introduced a related approach they called *chains of explanation*, which can be used to identify the causes of environmental degradation. As a methodological rule of the thumb, Blaikie and Brookfield (1987) recommend starting empirical examinations at the level of the immediate land manager, and work upwards in chains of explanations through national and global scales.

Methods

Land managers in chains of explanation can typically be smallholders. One may study how land and natural resources are used in practice by smallholders and ask what the reasons are for them using the resources in a particular way. And what are the social, economic and political conditions or forces that influence this land-use? These conditions can be found at various levels from the local to the global.

Figure 2.1 contains the main elements of a chain of explanations in a specific case. The direction of the arrows illustrates the stepwise process followed in such a study. Relations of power may, however, largely go in the opposite direction. But, chains of explanations may also be seen as rigid hierarchies of power, and in practice, power may move in different directions in what may be a web of relations rather than a chain (Rocheleau 2008). Nevertheless, having local sites as points of departure has often proven to be useful for identifying how various actors and processes influence and shape power relations through interaction between local and distant spaces across scales. Figure 2.2 gives an example of a web of relations to explain why elephants are killed by local people in West Kilimanjaro in Tanzania as a result of several factors working together (extension of protected areas, marginalization and dispossession of local people, growth of elephant populations, drought).

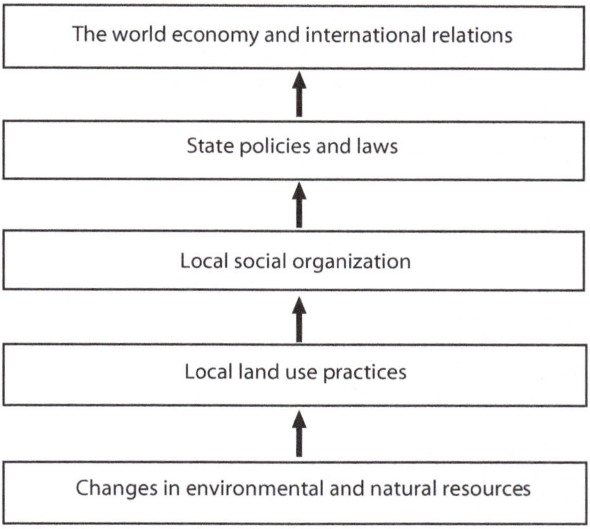

Fig. 2.1 The main elements in a chain of explanations. (Source: Created by the authors and inspired by Blaikie and Brookfield (1987))

2

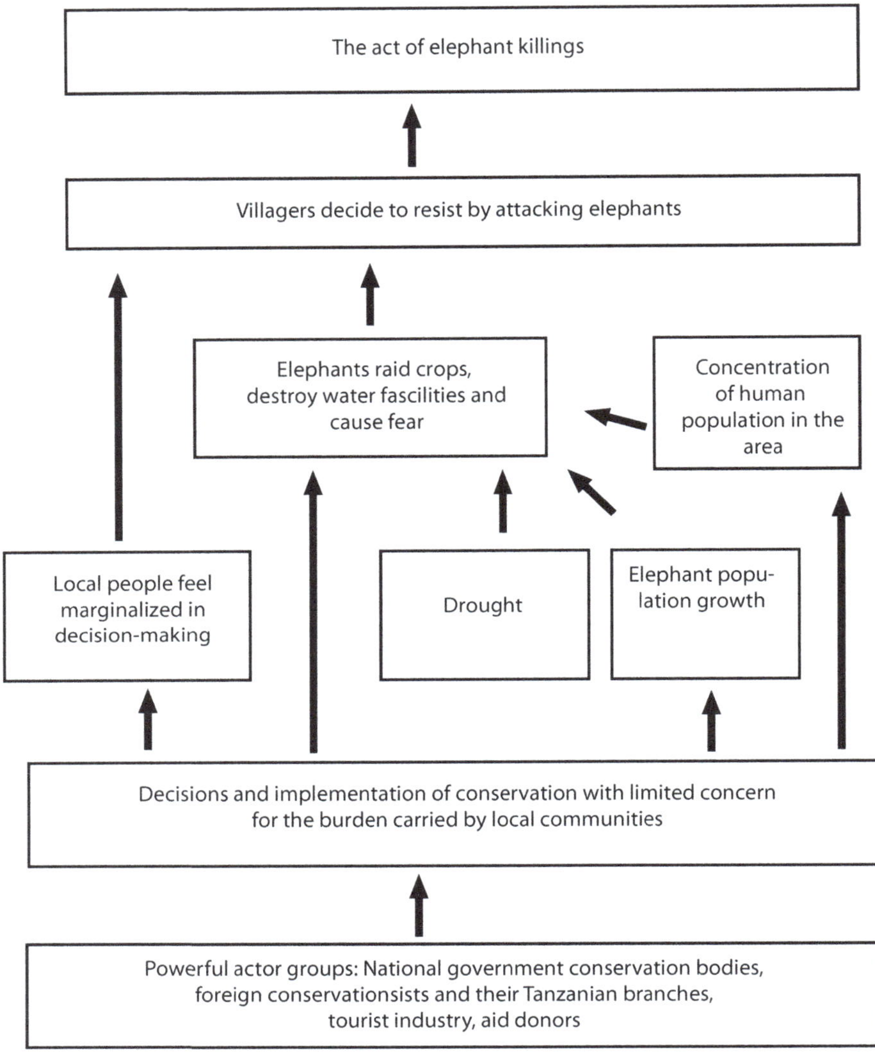

Fig. 2.2 A web of relations to explain elephant killings in a case in Tanzania. (Source: Adapted from Mariki et al. 2015)

A web of relations approach may also be seen as having some similarities with actor-network theory (Latour 2005) and its extension called 'assemblage' theory that has also been occasionally used within political ecology the last 10–15 years (Lave 2015). In such an approach, one studies networks, which includes human as well as non-human elements. Similarly, to human and cultural ecology, this approach may, however, be criticized for a weak conception of power, and even for neglecting power as a key issue (Lave 2015).

Vayda and Walters (1999) have, however, responded to the critique from political ecologists that human and cultural ecology lacks a focus on power and politics more broadly, and thereby fails to properly describe causes of environmental change. They argue that it is problematic that political ecology starts with a priori judgements giving privilege to certain political-economic explanations. As an alternative to political ecology, they introduce 'event ecology'.

This means

> to begin research with a focus on the environmental events or changes that we want to explain and then to work backward in time and outward in space so as to enable us to construct chains of causes and effects leading to those events or changes. By contrast, the practice of many political ecologists, presumably regarding access to resources as always politically determined and as always important for understanding or explaining environmental change, is to focus their research on such access, or on change in such access, and to pay little or not attention to actually demonstrating environmental effects (Vayda and Walters 1999: 169)

Hence, research should be 'guided more by open questions about why events occur than by restricted questions about how they are affected by factors privileged in advance by the investigator' (Vayda and Walters 1999: 170).

While event ecology seems to be another name for progressive contextualization, and while it is easy to agree that research should be open to all possible explanations, this critique may be seen to be relevant for only a section of political ecology scholarship, even in the late 1990s. Many political ecologists have after all engaged with ecology and not only relied on political explanations. One example is the contribution by political ecologists to the paradigm shift in drylands research and range science that took place in the 1990s (Stott and Sullivan 2000; Davis 2016; Sayre 2017). For instance, political ecology research in the African Sahel from the 1990s included climatic variation and non-equilibrium ecology as key elements in the analysis (Turner 1993 and Turner 1998, Benjaminsen 1997, Bassett and Bi Zuéli 2000, see also ► Chaps. 7 and 8). This research has questioned claims of desertification caused by overgrazing and has instead pointed at rainfall variation as the key driver of environmental change in the Sahel.

In addition, Piers Blaikie's book *The Political Economy of Soil Erosion in Developing Countries* from 1985 had paved the way for an approach to political ecology that engages explicitly with natural science (Blaikie 1985). The book was a critique of environmental conservation policies in the Global South and presented three central arguments (Neumann 2008). First, there is often lack of sound scientific data on soil erosion and other environmental processes, which leads to a high level of uncertainty. Second, actors perceive environmental change differently depending on their 'ideology'. Blaikie argued 'that *all* approaches to soil erosion and conservation are ideological – they are underpinned by a definite set of assumptions, both normative and empirical, about social change' (Blaikie 1985: 149). Third, environmental policies are about control over and rights to land and natural resources. A critical question that political ecology asks, therefore is: Who wins and who loses from resource and conservation policies? This again leads us to study 'where power lies and how it is used' (Blaikie 1985: 6).

Consequently, Piers Blaikie as a social scientist proposed an approach to understanding environmental problems by, on the one hand, problematizing the quality and uncertainty of scientific data, and on the other hand, insisting that the production, interpretation and use of environmental data are inherently political (Rigg 2006; Simon 2008).

In fact, Blaikie's two classics in political ecology (Blaikie 1985; Blaikie and Brookfield 1987) have inspired a whole stream of political ecology scholarship that engages with ecology and natural science, but without making a priori judgements about political-economic causation.

Despite this literature, also within political ecology itself, there has been a debate about the place and role of ecology. In a key contribution to this debate, Walker (2005) asks where the ecology is in political ecology and whether the field has become 'politics without ecology'.

In opposition to this view, Michael Watts has criticized some political ecology for paying too close attention to natural aspects ('ecology') and too little attention to 'political' aspects (Peet and Watts 1996; Watts 1997). Watts holds that this leads to an atheoretical approach, which lacks a general theory of social change that would explain environmental degradation. This debate may be seen to reflect different directions within political ecology between an approach engaging actively with natural science and ecology and one focusing on social theory and 'politics'.

In a contribution to this debate and echoing Vayda and Walters (1999), Forsyth and Walker (2008) argue that conventional approaches to political ecology have a priori linked environmental degradation to the impacts of global capitalism.

A critical political ecology would, however, critically and empirically examine all environmental representations whether based on Malthusianism or a critique of capitalism. This also implies investigating rather than assuming 'the essentialist link between capitalism and environmental degradation' that often is found in the development literature (Forsyth 2003). Critical political ecologies would combine deconstructions of narratives with a realist belief in science as a means to achieve more accurate descriptions and understandings of environmental realities. Such combinations of realist and constructivist positions are referred to as a critical realist position (see ► Chap. 1). According to Forsyth (2001), critical realism seeks to understand ecological change through a combination of epistemological scepticism and ontological realism.

Blaikie (1999), however, points out that political ecological critiques of claims of degradation in fact owe more to realist science than to constructionism. Hence, critical political ecology has to a large extent been based on realist investigations to identify environmental change and its causes.

2.3 Poststructuralism

In political ecology, Michel Foucault (1926–1984) has been the most influential poststructuralist thinker. Poststructuralism emerged in France in the 1960s as a further development of structuralism, which is a theory in linguistics associated primarily with the writings of the Swiss linguist Ferdinand de Saussure. Hence, both approaches have a focus on language. People do not only create language, but are also created by it. These labels ('structuralist', 'poststructuralist') have, however, rarely been used by scholars who have been associated with these ideas and several of the leading thinkers in the field can be seen as representatives of both structuralism and poststructuralism (e.g. Michel Foucault and Roland Barthes).

De Saussure had developed the idea that a sign can be seen as both 'signifier' (the form a sign takes) and 'signified' (the concept it refers to). Furthermore, signs are given meaning through the relationship with other signs within a structure (e.g. language). Poststructuralist writings, however, question this binary opposition between signifier and signified. Therefore, to understand an object it is necessary both to study the object itself as well as its meaning (i.e. the discursive level).

In practice, this has meant that poststructuralist writings have focused on what is said and written about an object or a phenomenon. Through historical studies on topics such as the medical treatment of madness and punishment and prison administration, Foucault (1979, 1988) showed how practice in different periods has been linked to discourses that have defined what has been seen as meaningful, true and accepted forms of statements and actions.

A discourse can be defined as an established knowledge regime linked to a particular phenomenon or theme (see ▶ Chap. 3 for more on discourses). Leading or hegemonic discourses are shared among powerful actors. One dimension of globalization is that a limited number of discourses dominate global debates. In this book, we will show different examples of how global discourses about environment and development can impact both national and local levels.

In environmental discourse analysis, nature is studied as a contested notion with the basic assumption that language not only reflects, but also shapes our views of the world and reality (Hajer and Versteeg 2005). Various discursive elements have been analysed since the emergence of environmental discourse analysis in the early 1990s with particularly narratives and metaphors as rhetorical devices occupying a central role (Hajer 1995; Dryzek 1997; Adger et al. 2001).

Images (drawings, photographs, maps) as forms of non-verbal rhetorical devices have been studied to a lesser extent in political ecology and environmental discourse analysis in general (Benjaminsen 2020). The recent focus on 'Nature 2.0' within political ecology, which is concerned with a combination of visual and textual analysis, may, however, be seen as an exception. Nature 2.0 refers to the increased use of social media by environmental organizations, which has rapidly changed the political economy of conservation, and led to a reimaging of nature and human-nature relations (Büscher 2016). This literature focuses on the study of human alienation from nature combined with remote environmental engagement through the use of new media, such as conservation websites, blogs, videos, photos, text and online chat forums. For instance, Igoe (2017) studies the use of romanticized and spectacular online photographs and videos of African wildlife and wilderness to sell safari destinations as well as conservation interventions, but which adversely result in alienation and commodity fetishism.

In discourse analysis carried out within environment and development studies as well as in political ecology, Emery Roe's approach to narrative analysis with a focus on crisis narratives has been influential (Roe 1991, 1999). Roe is especially concerned with arguments and courses of action in a narrative. In addition to such storylines, narratives also tend to contain a gallery of actors, such as the archetypical heroes, villains and victims (Svarstad 2009).

Narratives are stories with a beginning, middle and an end, or when cast in the form of an argument, with premises and conclusions. According to Roe (1999: 6),

> crisis narratives are the primary means whereby development experts, and the institutions for which they work, claim rights to stewardship over land and resources they do not own. By generating and appealing to crisis narratives, technical experts and managers assert rights as "stakeholders" in the land and resources they say are under crisis.

In a development context, this means that crisis narratives have portrayed marginalized people as the villains (and usually also the victims of their own mismanagement), while external experts associated with state polices are presented as the heroes who provide the solutions. In this way, states and other powerful actors have used such narratives to assert authority over land and natural resources at the expense of customary land-users (e.g. Leach and Mearns 1996; Stott and Sullivan 2000; Bassett and Bi Zuéli 2000; Adger et al. 2001).

Metaphors are also frequently used in environmental discourse (Larson 2014), as well as in the communication of crisis narratives. Well-known examples are 'spaceship earth', 'the sinking ark', 'greenhouse', population bomb', 'tragedy of the commons', 'land grabbing', 'carbon footprint' and 'creeping desert' to mention just a few. Metaphors are essential ingredients in communication and the rhetorical power of metaphors is well-known in politics and, in particular, by politicians and policymakers (Charteris-Black 2011).

However, as mentioned, the role of images in environmental discourse and in crisis narratives seems to be a neglected topic. To help us conceptualize the impact of images within environmental narratives, the classic work of Roland Barthes may be useful. Building on structuralist theory, Barthes (1961) articulates visual communication into the two separate levels of *denotation* (literal meaning of an image) and *connotation* (societal values communicated through an image). He added that image and text are closely connected, and that image does not necessarily illustrate text, but that it may often be the other way around. The caption, in particular, is included in the image denotation.

Later, Barthes (1972) added *myth* as a third layer, which refers to historically grounded ideological ideas that are evoked by an image. Myth is depoliticized speech that 'has the task of giving a historical intention a natural justification, and making contingency appear eternal' (Barthes 1972: 254).

Barthes' conception of 'myth' can be seen as closely linked to Foucault's idea about questioning what is taken for granted in a society. This has inspired political ecologists to study and question established 'truths' about various environment and development issues (Leach and Mearns 1996; Stott and Sullivan 2000; Adger et al. 2001; Paulson and Gezon 2005). Such truths have in political ecology been studied as mainstream views, received wisdom, dominant or policy narratives and myths.

However, it will often be difficult to separate in practice these three levels of meaning, because interpretation will vary with cultural background and also

Photo 2.1 Cracked clay is an image often used to depict desertification. This is usually soil which cracks after having been flooded. Areas with clay are usually found in the lower parts of landscapes and tend to be flooded in the rainy season before they dry up in the dry season. When searching photo data bases for 'desertification', various similar photos of cracked clay are offered. Especially around the annual World Day to Combat Desertification (17th June), such images accompany articles on desertification published on websites and in printed media on a global scale. (*Source: Getty Images/Athul Krishnan*)

because the levels may overlap to some degree (Chandler 2017). Nevertheless, the African desert landscape in Photo 2.1 may tentatively serve as an example. The denotative meaning of this image would simply be a dry landscape with cracked soil, and the connotative meaning could refer to nature under threat through overexploitation. The mythical meaning would be related to the historically manufactured idea of 'desertification' (Benjaminsen and Berge 2004).

This is an old colonial trope originally linked to European perceptions of environmental degradation, such as overgrazing, deforestation, soil erosion and general careless mismanagement by African smallholders. Photo 2.1 is an example of an iconic image that efficiently transmits ideas of desertification conveying variegated associations to its viewers depending on their knowledge and background. Yet, among most observers, the photograph will generate the idea of desert encroachment (in Africa) caused by local mismanagement or climate change, or both in combination. This is also an idea that tends to resist empirical evidence, such as the fact that the drylands in the Sahel have become greener during the last 30 years (Benjaminsen and Hiernaux 2019).

Myths move beyond connotation, because they are taken for granted and not questioned. In this way, myths have much in common with crisis narratives that also resist empirical knowledge that goes against their storyline (Roe 1999). As Barthes points out, myths are depoliticized stories that work to conceal the ideological function of images. In Photo 2.1, the hidden ideology would be a Eurocentric, Malthusian and neo-colonial view of African landscapes and their users, thereby contributing to the justification of the use of strict measures to arrest desertification—through fines, restricted use and sometimes the use of violence.

While discourses are structures that are produced, reproduced and changed by actors, Foucault has been criticized for a conception of what some see as actor-less structures (Fox 1998). People are, however, rarely passive victims of the mechanisms of discourses, but act within the possibilities that discourses and material circumstances provide (Stern et al. 2015).

Definition

Foucault's notion of *governmentality* has increasingly been used in political ecology as a way to understand how power works in environmental governance (e.g. Agrawal 2005; Fletcher 2010, 2017; Bose et al. 2012; Valdivia 2015; Johnsen and Benjaminsen 2017). Governmentality can be seen as ways in which governments administer citizens through discourse to act in accordance with government priorities (Foucault 1991, 2008). Buliding an Foucault (2008), and as mentioned in ► Chap. 1, Fletcher (2010) distinguishes between four different governmentalities that are relevant to environmental governance. 'Disciplining' implies that the government manages to get citizens to internalize certain 'mentalities' in terms of social norms and ethical standards. 'Truth' means ruling of people through religion or other overarching truth-defining principles. The other two types of governmentality are not necessarily dependent on people subscribing to believing state presentations or priorities, although they will see it as beneficial to act in accordance with them. 'Neoliberal rationality' implies that an incentive structure is established to maximize results, while 'sovereign power' means governing through defined rules and sanctions. Each of these governmentalities may work alone, overlap or conflict with any of the other forms (Fletcher 2010). The four techniques may sometimes function separately, sometimes work together and they may also be in conflict. These are power techniques that not only states may use, but also actor groups such as universities, political parties, and companies (Li 2007).

There are also other perspectives, influenced by Foucault, that explicitly allow more space for agency. For instance, in his 'third power dimension', Lukes (2005) deals with discursive power where one actor gets other actors to do something they would otherwise not do, by influencing their wishes. As mentioned in ► Chap. 1, this may involve governments perspectives domination media and education, so that people only get access to perspectives in line with those within government. In this situation, people are not forced to act in the manner that the government wants, but they choose to do it themselves based on the information they have access to.

Furthermore, Edward Said (1935–2003) has inspired political ecology, especially through his renowned book on 'Orientalism' (Said 1978), which largely also inspired the emergence of postcolonial studies as a new field. Said saw Orientalism as a discourse produced in the West about the inferior 'others' in 'the Orient'. Western fiction, visual art, academia and journalism have consistently contributed to this discourse presenting the 'Orient' as alien, different and inferior to Western culture. In this way, 'Orientalism' has helped to justify colonialism and Western domination, Edward Said argued. According to this thinking, Western cultural expressions can be seen more as reflections of the interests of Western actors, than as presentations of reality in the 'Orient'.

Said was also influenced by Foucault and believed that all statements are fundamentally ideological, especially when they concern other cultures. While he mainly discussed the relationship between the West and the Arab countries, inspired by Orientalism recent postcolonial studies have focused on other parts of the world, such as Africa, Asia and the Arctic.

Definition

Furthermore, Foucault's concept of *biopower* has also been used to some extent within political ecology (Cavanagh 2018). Foucault distinguishes between sovereign power as the power to 'take life or let live' and biopower as the power to 'make live or let die'. Biopower implies that in order to secure lives, governmental concerns have emerged about the qualities of various populations such as health, and opportunities for improvement of living standards. Foucault opens up perspectives on how disciplinary forms of power-knowledge have come to normatively prescribe how both individuals and populations should behave. Moreover, as people are 'intricated' in a 'perpetual conjunction' with nature, the sovereign will have to intervene by acting on the environment if he wants to change human populations. Thus, biopolitical forms of knowledge and governance perceive the environment as a central object of concern (Cavanagh 2018).

In political ecology, the notion of biopower is reflected in studies of famine, food security, infectious diseases and climate change adaptation that critically examine the power–knowledge mechanisms involved when the justification for security interventions are based on unclear distinctions between the wellbeing of humans and the environment. Biopower is, however, a perspective that has been examined only to a limited degree so far by political ecologists (Cavanagh 2018).

2.4 Peasant Studies

The roots of peasant studies lie in efforts to explain the motivations and behaviour of peasants who may be seen as pastoralists and small-scale farmers. A key question has for instance been to understand why peasants tend to resist modernization. In parallel to Netting's cultural ecology, such agrarian studies often imply a critique of dominant ideas about peasants being irrational and inefficient producers.

James Scott, who is professor of agrarian studies at Yale University, is one of the leading scholars in peasant studies. In *The Moral Economy of the Peasants: Rebellion and Subsistence in Southeast Asia* (1976), Scott argued that peasants have a tendency to attempt to avoid risk by developing social institutions for the redistribution of surplus in order to protect themselves against bad years. Such arrangements may include the communal sharing of land and labour.

In *Weapons of the Weak: Everyday Forms of Peasant Resistance* (1985), Scott described what he called the peasants' *everyday resistance* to modernization and exploitation. Such tactics of non-compliance require little or no coordination or planning and include 'foot dragging, dissimulation, false compliance, pilfering, feigned ignorance, slander, arson, sabotage, and so forth' (Scott 1985: 29). Scott refers to these everyday acts of resistance as the 'weapons of the weak'; the 'weak' being actors who use a variety of 'weapons' to defend their interests against more powerful and dominating actors. In some cases, resistance can be more effective when hidden than open, because direct confrontation may provoke a response from the elite that could lead to further loss to communities.

Scott argues, however, that though the acts of hidden resistance do not openly 'contest the formal definitions of hierarchy and power', it is possible to determine to what degree, and in what ways, marginalized actors accept the social order propagated by elites by studying their behaviour and 'offstage' comments and conversations.

The 'offstage' presentations, or 'hidden transcripts' as Scott (1990) calls them, are accounts that include critiques of power and practices. Hidden transcripts can also be disguised in the form of rumours, proverbs, jokes, parodies, gossip, gestures, folktales, and so on. 'Public transcripts', on the other hand, are comments and conversations that the actors (the elite and the inferior) play out in each other's presence (Scott 1990). Scott (1990: 2) explains that while public transcripts can inform us about power, they are 'unlikely to tell the whole story about power relations'. Therefore, he recommends examining the divergence between the public and hidden transcripts as an approach to study structures of domination and patterns of resistance.

While peasants are partially integrated in markets and therefore part of the capitalist system, a peasant unit is still not a fully capitalist unit of production, because it operates according to its own complex logic involving both market and non-market incentives for or drivers of land use change (van der Ploeg 2013). Peasant labour, for instance, is often derived from family members rather than provided in exchange for a wage. Similarly, in peasant communities, capital primarily consists of available tools, buildings, livestock and savings rather than necessarily as reinvested surplus-value producing capital as theorized by Marxist scholars.

For these reasons, and linked to what the Russian agricultural economist Alexander Chayanov (1888–1937) termed 'self-exploitation' (Chayanov 1925), peasant farms tend to be more competitive than capitalist farms. When capitalist farms go bankrupt, peasants in the same areas may work longer hours and sell at lower prices (van der Ploeg 2013). Peasants may also be able to use more marginal land than capitalist farms, and they use land more intensely, adding especially more labour to each unit of land, which can lead to higher yields.

More recently, and as a consequence of the fact that contemporary agrarian questions involve more actors than only peasants, critical agrarian studies have emerged out of peasant studies as a field focusing on how agrarian life and livelihoods shape and are shaped by the politics, economics and social worlds of modernity (Edelman and Wolford 2017). Marginalization, marginality and the impact of modernity on marginal people constitute key themes in critical agrarian studies (Edelman and Wolford 2017).

Definition

Political ecology is a critical approach that also builds on the legacy of *critical theory* from the Frankfurt School. Max Horkheimer stated that traditional theory was oriented towards understanding and explaining aspects of society, while critical theory also has a liberating dimension by pointing out situations that should be changed. In order to achieve this ambition, Horkheimer believed that a broad interdisciplinary approach was necessary (Horkheimer 1970). During the 1960s, Jürgen Habermas and other scholars of the Frankfurt School criticized positivism for its claims of objectivity having a weak epistemological foundation. Habermas' alternative was anchored in a critical philosophy of science, where natural, cultural and social science were linked to their technical, practical and liberating interests (Habermas 1968). The impact on political ecology seems evident here, although curiously seldom made explicit in political ecology scholarship. One of the most influential books in the field is for instance called 'Liberation Ecologies' (Peet and Watts 1996), but without including a reference to the Frankfurt School, its critical theory and focus on *liberating* interests.

Political ecology has mostly provided critical studies pointing out what there is a need to be liberated *from*. However, political ecology has thus far played a limited role in clarifying what this liberation should *lead to*. This reflects the fact that political ecology's political influence remains modest, and that apolitical ecology has obtained much more political influence than political ecology. However, political ecology-based critique may also include ideas for transformative change (Cavanagh and Benjaminsen 2017), while apolitical ecology tends to result in small practical solutions within frames that are not problematized.

2.5 The Interface Between Political Ecology and Environmental Justice

Environmental Justice (EJ) is a related scholarly field that has been more closely linked to activism for social justice than political ecology (PE). As scholarly approaches, EJ and PE have many parallels through common normative aspirations and a focus on justice (Svarstad and Benjaminsen 2020). Thematically, EJ and PE overlap in that they both involve critical studies of environmental interventions.

EJ emerged in the USA in the 1970s and 1980s in the form of civil rights struggles against the dumping of hazardous waste and was closely linked to issues of race and class (Schlosberg 2004; Walker 2012; Agyeman 2013; Holifield et al. 2018). The earliest contributions to academic scholarship on EJ came from sociologists such as Robert Bullard, who published the article *Solid waste sites and the black Houston community* in 1983, and Stella Capek (1993) who contributed an early social constructivist perspective on EJ struggles.

Since the 1990s, scholars from various disciplines have contributed to the development of EJ in the USA. For instance, the geographer Laura Pulido (1996) studied resistance among Mexican Americans caused by exposure to pesticides and conflicts over grazing. Her analysis focused on political economic structures and the role of racism in causing environmental injustice. In a similar vein, the sociologist David Pellow (2002) analysed the politics of garbage dumping in Chicago and showed how minority and poor communities carry the costs of this dumping in the form of exposure to health risks. He also demonstrated the role of class and race in the production of these outcomes.

Since the turn of the millennium, EJ scholarship has expanded geographically to places other than the USA. In addition, this literature has widened, both thematically and in terms of academic framing (Martinez-Alier et al. 2014; Holifield 2015; Holifield et al. 2018). As a result, EJ perspectives are now being used on global scales (Agyeman 2013; Martin et al. 2013; Mehta et al. 2014; Martinez-Alier et al. 2016), such as in the Global South (Schroeder et al. 2008; Sikor and Newell 2014; Martin et al. 2016), and in other geographical locations such as Europe, Australia, and the Arctic (Köckler et al. 2018; Steger et al. 2018; Schlosberg et al. 2018; Shaw 2018). This expansion has, however, generated a critique that EJ tends to employ a Western and universalist analytical framework that results in further domination and misrecognition (Lawhon 2013; Martin et al. 2013; Vermeylen 2019). Hence, EJ may yet need to be decolonized (Álvarez and Coolsaet 2018; Fraser 2018; Temper 2019).

While EJ emerged with a focus on pollution in the USA, PE, as mentioned in ▶ Chap. 1, initially concentrated on the governance of renewable natural resources (such as soils, forests, pastures) in the Global South. In the last decade or two, PE scholarship has been expanded to urban political ecology (Swyngedouw and Heynen 2003; Holifield 2009; Heynen 2014), and environments in the Global North (McCarthy 2005; Schroeder et al. 2006; Benjaminsen and Robbins 2015). Although it has been claimed that PE also needs to be decolonized (Schulz 2017), PE scholars have been at the forefront of decolonizing environmental science (Neimark et al. 2019), exploring themes such as 'conservation' (Adams and Mulligan 2003; Salazar Parrenas 2018), 'development' (Wainwright 2008), 'environmental education' (Meek and Lloro-Bidart 2017), and 'food justice' (Bradley and Herrera 2016).

Although a number of authors mention both EJ and PE and note that they overlap, there have been few explicit comparisons or discussions of the two fields together. Exceptions are Holifield (2009, 2015) and Ranganathan and Balazs (2015), although limited to a discussion of EJ and urban political ecology in addition to our own reading of EJ through a PE lens (Svarstad and Benjaminsen 2020). In his analysis of the overlaps between the two bodies of work, Holifield (2015: 585) points out that EJ and PE have been 'traveling down quite different paths' and only recently started to meet, despite seemingly being such 'a perfect match'.

Holifield (2015) also points out considerable differences between mainstream EJ and PE. Methodologically, EJ has traditionally been grounded in quantitative descriptions, while PE has been dominated by qualitative methods and case studies, although there are exceptions to these trends in both fields. The same author refers to critique of mainstream EJ research by some political ecologists (Swyngedouw and Heynen 2003) as lacking a focus on (radical) theory. When mainstream EJ has been theorized, it has been dominated by rational choice and systems models, which according to Holifield (2015), explains why EJ and PE have remained separate for so long.

Following Cook and Swyngedouw (2012), Holifield (2015) comments that mainstream EJ focuses primarily on patterns of socio-environmental inequalities, while PE is concerned with explaining the production of such inequalities. Nevertheless, Holifield recognizes contributions from critical and radical scholars from the early history of EJ, noting that by the turn of the millennium, such contributions 'had set the stage for deeper engagements between political ecology and alternative approaches to environmental justice' (Holifield 2015: 589).

Ranganathan and Balazs (2015) demonstrate the fruitfulness of drawing from a blend of EJ and urban political ecology perspectives in their comparative analysis of two cases of water marginalization across the Global North/South divide. They include within EJ scholarship what they see as health-oriented, as well as critical theoretical strands, combining this with urban political ecology based on Marxist approaches, post-colonialism, and with an explicit focus on power relations regarding access to clean and safe water.

Definition

Today, the dominant approach within EJ is what we call a *radical EJ approach* (Svarstad and Benjaminsen 2020), which emerged when Schlosberg (2004) adopted the three elements of *distribution*, *recognition* and *procedural justice* from the field of political philosophy, in particular the work of Nancy Fraser (1998, 1999), but also that of Iris Young (1990) and Axel Honneth (1995, 2001). Nancy Fraser initially proposed a two-dimensional focus on *distribution* and *recognition*, but in the second half of the 1990s she included *representation* as a third dimension (Fraser 1998, 1999), while Young (1990) had already included these three dimensions along with others.

With a focus on the three elements of distributive justice, recognition and procedural justice, radical EJ has evolved rapidly into a sort of leading template for many EJ studies. It was later expanded in various ways by Schlosberg and other scholars, notably by adding a radical version of *capabilities theory* as a fourth element (Schlosberg 2007; Walker 2012; Sikor 2013).

The importance of pluralistic approaches is often emphasized within the radical EJ framework (Schlosberg 2004, 2007). Reflecting on these developments, Holifield et al. (2018: 4) observe that

> … scholars have increasingly addressed and theorized the multidimensionality of the justice in environmental justice. In addition to *distributive* justice, which remains a key focus of much quantitative and spatial analysis, a growing body of literature now attends to *procedural* and participatory justice, justice as *recognition*, and justice as *capabilities*, as well as the interrelations among these dimensions.

Distributive justice refers to the distribution of burdens and benefits related to environmental interventions. *Justice as recognition* concerns who is given respect (or not) and whose interests, values and views are recognized and taken into account. *Procedural justice* is about who is involved and has influence in terms of decision-making, while *capabilities theory* focuses on the extent to which people are able to live the lives they consider to be valuable.

As pointed out by Fraser (2018), a considerable part of early EJ scholarship drew on the work of the liberal philosopher John Rawls (1999). However, this liberal justice tradition has to some extent been abandoned within EJ, particularly so in the literature on radical EJ. As described above, the first three elements of radical EJ were drawn from radical justice philosophy that contrasts with the liberal justice perspectives of Rawls. On the other hand, the fourth element of the radical EJ framework (capabilities theory) is often characterized as a liberal theory—especially in Sen's version—due to its focus on individual choices and deliberate processes. Schlosberg, however, applies what Edwards et al. (2016) interpret as a radical version of this theory, with a primary focus not on the hedonic well-being of individuals, but on the functioning of communities and collective goods. Thus, all four elements of the radical EJ framework may be seen as constituting a radical alternative to liberal EJ theory.

The literature that builds on the radical EJ framework has become extensive over the last decade. Examples of the broad range of EJ topics discussed are: elderly people in urban neighbourhoods (Day 2010); ecosystem services (Sikor 2013); biodiversity conservation (Martin et al. 2013; Lecuyer et al. 2018); climate change interventions in cities (Bulkeley et al. 2014); and carbon offset forestry projects (Fisher et al. 2018).

Walker (2012) also uses this radical EJ approach in his expanded framework focusing on the main distinctions between claims about justice, evidence and process. In Walker's framework, all such claims can be classified as being about distribution, recognition, procedural justice or capabilities.

Radical EJ is, however, only one of the various types of EJ that may be applied today. For instance, contributions are still made to liberal EJ studies. In addition, there is an approach known as 'critical EJ' (Pellow and Brulle 2005; Pellow 2016, 2018), which—with its multi-scalar approach and critique of the state—has much in common with PE. Critical EJ is based on four pillars (Pellow 2018: 17–18): (1) Paying more attention to multiple social categories; (2) Promoting multi-scalar methodological and theoretical approaches; (3) Pursuing transformative rather than reformist approaches to challenge inequality and resist the power of the state; and (4) Focusing intensely on how humans and more-than-human nature are indispensable in terms of building just sustainabilities.

Furthermore, Pulido and De Lara (2018) have recently argued for yet another alternative way to frame EJ, based on abolitionist theories and decolonial epistemologies, which may serve to critique power relations without depending on state recognition. Agreeing with Pellow, Pulido and De Lara view the lack of power analyses as problematic aspects in mainstream EJ (including radical EJ).

2.6 *Écologie Politique, Ecología Política* and Degrowth

The political ecology we present in this book refers mainly to the approach that emerged in the USA and the UK in the 1980s. But there is also a southern European version of political ecology (*écologie politique*, *ecología política*), which again is closely connected to the degrowth movement that started in France in the 1970s (known as *décroissance*). During the last few years, there has been increasing interest in degrowth across Europe in particular where it has grown to a movement of considerable size with large biennial conferences (D'Alisa et al. 2014). According to Kallis (2017: 10) degrowth is 'an equitable down-scaling of production and consumption that increases human well-being and enhances environmental conditions'.

The southern European version of political ecology is more closely linked to activism and socio-environmental transformation with closer associations and engagements with green political parties and not the least with environmental justice principles and organizations (Martinez-Alier 2002). For instance, in France *écologie politique* refers to 'green or ecological politics' and to the ideology, politics and advocacy of the green party (*Les Verts*). As a consequence, a textbook about 'mainstream' English-language-based political ecology in French had to use the English term (Gautier and Benjaminsen 2012) (see also Benjaminsen and Svarstad 2009; Chartier and Rodary 2015).

The southern European version of political ecology is also less deconstructionist and less critical of claims about 'degradation', 'scarcity' and 'limits'. For instance, the publication of the report about *Limits to Growth* by the Club of Rome (Meadows et al. 1972) was instrumental in sparking the emergence of both versions of political ecology, but in different ways; in the Anglophone version it led to a critique of Malthusianism and its lack of social or class analysis in its presentation of environmental crisis scenarios, but in France the report inspired the emergence of *décroissance* and *écologie politique*.

In general, southern European political ecology seems to have had a less problematic relationship to Malthusianism than what has been the case in 'mainstream' English-language-based political ecology, although there are exceptions to this (Martinez-Alier 2002; Kallis 2019).

Kallis (2019) defends degrowth and argues for 'self-limitation', which is different from a Malthusian view of limits. He argues that degrowth supporters like him do not call for limits because the world is running out of resources, but because they want to end growth in order to stop ecological and social destruction.

A recent debate between political ecologist Paul Robbins (2020b) and degrowth supporter Erik Gómez-Baggethun (2020) can also be seen to reflect contrasting positions across this divide between the two directions of political ecology, although Robbins' modernist eco-socialist position may be contested also among 'mainstream' political ecologists. Robbins points out that political ecology and degrowth has many commonalities and are compatible to some extent. But there are also fundamental differences, it seems:

> » First and foremost, political ecology continuously and convincingly has made strong arguments against *natural limits*. Whether deconstructing carrying capacity …, undermining the inanity of Malthusianism …, or challenging the political assumptions of planetary boundaries …, political ecological stories have always pointed to the way scarcity is a construct that is allied with elite power, not emancipatory process. Degrowth's insistence on the urgency of less, as a discourse, puts it in friction with political ecology's rejection of elitist arguments and policies that advocate forms of self-denial.' (Robbins 2020b: 3)

2

Gómez-Baggethun (2020: 2) responds first by stating that while limits may be flexible and to some extent constructed, 'the economy cannot grow perpetually: the scale of the economic sub-system is limited by the size of the host ecosystem', and 'in today's post-truth era, banalizing research on ecological limits as mere narratives or social constructs pays service (albeit unintendedly) to the same elites and business powers against which such claims where initially conceived'.

With the growing threat and urgency of climate change, political ecology becomes increasingly torn between on the one hand deconstructions of claims and narratives of degradation, limits and scarcity emerging from environmental science and on the other hand contributing to alternative sustainabilities within the operating space of planetary boundaries.

The two political ecology traditions have, however, recently found a meeting and discussion space in the international network on political ecology called POLLEN (► politicalecologynetwork.org) including the organization of biennial conferences where these questions are being further explored.

Questions

1. Identify and define at least two conceptual tools from Marxist political economy that are used in political ecology.
2. Select a case of an environmental conflict you know and try to identify causes through a chain of explanations or web or relations.
3. Give two examples of how poststructuralism has influenced political ecology.
4. What are the main influences of peasant studies on political ecology?
5. What are the similarities and differences between environmental justice and political ecology?
6. Do you see a tension in political ecology between deconstructions of degradation, limits and scarcity on the one hand and taking seriously notions of sustainability within planetary boundaries on the other hand? How can we deal with this tension?

References

Adams, W., and M. Mulligan, eds. 2003. *Decolonizing nature: Strategies for conservation in a post-colonial era*. London: Earthscan.

Adger, W.N., T.A. Benjaminsen, K. Brown, and H. Svarstad. 2001. Advancing a political ecology of global environmental discourses. *Development and Change* 32 (4): 681–715.

Agrawal, A. 2005. *Environmentality: Technologies of government and the making of subjects*. Durham: Duke University Press.

Agyeman, J. 2013. Global environmental justice or Le droit au monde? *Geoforum* 54: 236–238.

Álvarez, L., and B. Coolsaet. 2018. Decolonizing environmental justice studies: A Latin American perspective. *Capitalism, Nature, Socialism* 2018: 1–20. https://doi.org/10.1080/10455752.2018.1558272.

Barthes, R. 1961 (1977). *Image, music, text*. New York: Hill & Wang.

———. 1972 (2013). *Mythologies*. New York: Hill & Wang.

Bassett, T.J., and K. Bi Zuéli. 2000. Environmental discourses and the Ivorian savanna. *Annals of the Association of American Geographers* 90 (1): 67–95.

Bates, D.G., and J. Tucker, eds. 2010. *Human ecology: Contemporary research and practice*. New York: Springer.

Benjaminsen, T.A. 1997. Natural resource management, paradigm shifts and the decentralization reform in Mali. *Human Ecology* 25 (1): 121–143.

———. 2015. Political ecologies of environmental degradation and marginalization. In *The Routledge handbook of political ecology*, ed. T. Perreault, G. Bridge, and J. McCarthy, 354–365. London: Routledge.

———. 2020. Depicting decline: Images and myths in environmental discourse analysis. *Landscape Research*: 1–15. https://www.tandfonline.com/doi/full/10.1080/01426397.2020.1737663#.

Benjaminsen, T.A., and G. Berge. 2004. Myths of Timbuktu – From African El Dorado to desertification. *International Journal of Political Economy* 34 (1): 31–59.

Benjaminsen, T.A., and I. Bryceson. 2012. Conservation, green/blue grabbing and accumulation by dispossession in Tanzania. *Journal of Peasant Studies* 39 (2): 335–355.

Benjaminsen, T.A., and P. Hiernaux. 2019. From desiccation to global climate change: A history of the desertification narrative in the West African Sahel, 1900-2018. *Global Environment* 12: 206–236.

Benjaminsen, T.A., and P. Robbins. 2015. Nordic political ecologies. *Norwegian Journal of Geography* 69 (4): 1–6.

Benjaminsen, T.A., and H. Svarstad. 2009. Qu'est-ce que la 'political ecology'? *Natures, Sciences, Sociétés* 17 (1): 3–11.

Bergius, M., T.A. Benjaminsen, and M. Widgren. 2018. Green economy, Scandinavian investments and agricultural modernization in Tanzania. *Journal of Peasant Studies* 45 (4): 825–852.

Bergius, M., T.A. Benjaminsen, F. Maganga, and H. Buhaug. 2020. Green economy, degradation narratives and land-use conflicts in Tanzania. *World Development* 129: 104850.

Blaikie, P. 1985. *The political economy of soil erosion in developing countries*. New York: Longman.

———. 1999. A review of political ecology: Issues, epistemology and analytical narratives. *Zeitschrift für Wirtschaftsgeographie* 43 (3–4): 131–147.

Blaikie, P., and H. Brookfield, eds. 1987. *Land degradation and society*. London and New York: Methuen.

Bose, P., B. Arts, and H. van Dijk. 2012. 'Forest governmentality': A genealogy of subject-making of forest-dependent 'scheduled tribes' in India. *Land Use Policy* 29: 664–673.

Bradley, K., and H. Herrera. 2016. Decolonizing food justice: Naming, resisting, and researching colonizing forces in the movement. *Antipode* 48 (1): 97–114.

Brown, J.C., and M. Purcell. 2005. There's nothing inherent about scale: Political ecology, the local trap, and the politics of development in the Brazilian Amazon. *Geoforum* 36: 607–624.

Bulkeley, H., G.A.S. Edwards, and S. Fuller. 2014. Contesting climate justice in the city: Examining politics and practice in urban climate change experiments. *Global Environmental Change* 25: 31–40.

Bullard, R.D. 1983. Solid waste sites and the black Houston community. *Sociological Inquiry* 53: 273–288.

Büscher, B. 2009. Letters of gold: Enabling primitive accumulation through neoliberal conservation. *Human Geography* 2 (3): 91–94.

———. 2016. Nature 2.0: Exploring and theorizing the links between new media and nature conservation. *New Media & Society* 18 (5): 726–743.

Capek, S.M. 1993. The 'environmental justice' conceptual discussion and an application. *Social Problems* 40 (1): 5–24.

Cavanagh, C. 2018. Political ecologies of biopower. *Journal of Political Ecology* 25 (1): 402–425.

Cavanagh, C., and T.A. Benjaminsen. 2017. Political ecology, variegated green economies, and the foreclosure of alternative sustainabilities. *Journal of Political Ecology* 24: 200–216.

Cavanagh, C., T. Weldemichel, and T.A. Benjaminsen. 2020. Gentrifying the African landscape: The performance and powers of for-profit conservation on southern Kenya's conservancy frontier. *Annals of the American Association of Geographers* 110 (4): 1–19.

Chandler, D. 2017. Semiotics for beginners. http://visualmemory.co.uk/daniel/Documents/S4B/sem06.html.

Charteris-Black, J. 2011. *Politicians and rhetoric: The persuasive power of metaphor*. London: Palgrave Macmillan.

2

Chartier, D., and E. Rodary. 2015. French *écologie politique*: A political necessity. In *The international handbook of political ecology*, ed. R.L. Bryant, 547–560. Cheltenham: Edward Elgar.

Chayanov, A. 1925 (1966). *The theory of peasant economy*. Manchester: Manchester University Press.

Cohen, A., and K. Bakker. 2014. The eco-scalar fix: Rescaling environmental governance and the politics of ecological boundaries in Alberta, Canada. *Environment and Planning D: Society and Space* 32: 128–146.

Cook, I.R., and E. Swyngedouw. 2012. Cities, social cohesion and the environment: Towards a future research agenda. *Urban Studies* 49: 1959–1979.

Coquery-Vidrovitch, C. 1975. Research on an African mode of production. *Critique of Anthropology* 2 (4–5): 38–71.

Corson, C. 2011. Territorialization, enclosure and neoliberalism: Non-state influence in struggles over Madagascar's forests. *Journal of Peasant Studies* 38 (4): 703–726.

D'Alisa, G., F. Demaria, and G. Kallis, eds. 2014. *Degrowth: A vocabulary for a new era*. London: Routledge.

Davis, D.K. 2016. *The arid lands: History, power, knowledge*. Cambridge: MIT Press.

Day, R. 2010. Environmental justice and older age: Consideration of a qualitative neighbourhood-based study. *Environment and Planning A* 42: 2658–2673.

Dryzek, J.S. 1997. *The politics of the earth. Environmental discourses*. Oxford: Oxford University Press.

Edelman, M., and W. Wolford. 2017. Introduction: Critical agrarian studies in theory and practice. *Antipode* 49 (4): 959–976.

Edwards, G.A.S., L. Reid, and C. Hunter. 2016. Environmental justice, capabilities, and the theorization of well-being. *Progress in Human Geography* 40 (6): 754–769.

Ehrlich, P. 1968. *The population bomb*. New York: Ballantine Books.

Ekers, M., and S. Prudham. 2015. Towards the socio-ecological fix. *Environment and Planning A: Economy and Space* 47: 2438–2445.

Enzensberger, H.M. 1974. A critique of political ecology. *New Left Review* 8: 3–32.

Fisher, J.A., C. Cavanagh, T. Sikor, and D.M. Mwayafu. 2018. Linking notions of justice and project outcomes in carbon offset forestry projects: Insights from a comparative study in Uganda. *Land Use Policy* 73: 259–268.

Fletcher, R. 2010. Neoliberal environmentality: Towards a poststructuralist political ecology of the conservation debate. *Conservation and Society* 8 (3): 171–181.

———. 2017. Environmentality unbound: Multiple governmentalities in environmental politics. *Geoforum* 85: 311–315.

Forsyth, T. 2001. Critical realism and political ecology. In *After postmodernism: Critical realism?* ed. A. Stainer and G. Lopez. London: Athlone Press.

———. 2003. *Critical political ecology. The politics of environmental science*. London: Routledge.

Forsyth, T., and A. Walker. 2008. *Forest guardians, forest destroyers: The politics of environmental knowledge in northern Thailand*. Seattle: University of Washington Press.

Foster, J.B. 1999. Marx's theory of metabolic rift: Classical foundations for environmental sociology. *The American Journal of Sociology* 105 (2): 366–405.

Foucault, M. 1979. *Discipline and punish: The birth of the prison*. Harmondsworth: Penguin Books.

———. 1988. *Madness and civilization: A history of insanity in the age of reason*. New York: Vintage.

———. 1991. Governmentality. In *The Foucault effect: Studies in governmentality*, ed. G. Burchell, C. Gordon, and P. Miller. Chicago: University of Chicago Press.

———. 2008. The Birth of Biopolitics: lectures at the College de France 1978–1979. Newyork: Picador.

Fox, N.J. 1998. Foucault, Foucauldians and sociology. *British Journal of Sociology* 49 (3): 415–433.

Fraser, N. 1998. Social justice in the age of identity politics: Redistribution, recognition, and participation. *The Tanner lectures on human values*, Delivered at Stanford University, April 30–May 2, 1996.

———. 1999. Social justice in the age of identity politics: Redistribution, recognition and participation. In *Culture and economy after the cultural turn*, ed. L. Ray and A. Sayer. London: Sage.

Fraser, J. 2018. Amazonian struggles for recognition. *Transactions of the Institute of British Geography* 43: 718–732.

Gautier, D., and T.A. Benjaminsen, eds. 2012. *Environnement, discours et pouvoir. L'approche Political Ecology*. Versailles: Quae.

Gómez-Baggethun, E. 2020. More is more: Scaling political ecology within limits. *Political Geography* 76: 102095.

Habermas, J. 1968. *Technik und Wissenschaft als 'Ideologie'*. Frankfurt am Main: Suhrkamp Verlag.

Hajer, M. 1995. *The politics of environmental discourse*. Oxford: Oxford University Press.

Hajer, M., and W. Versteeg. 2005. A decade of discourse analysis of environmental politics: Achievements, challenges, perspectives. *Journal of Environmental Policy & Planning* 7 (3): 175–184.

Hall, D., P. Hirsch, and T.M. Li. 2011. *Powers of exclusion. Land dilemmas in Southeast Asia*. Singapore: National University of Singapore Press.

Hardin, G. 1968. The tragedy of the commons. *Science* 162: 1243–1248.

Harvey, D. 1999. *The limits to capital*. London: Verso.

———. 2003. *The new imperialism*. Oxford: Oxford University Press.

———. 2014. *Seventeen contradictions and the end of capitalism*. London: Profile Books Ltd.

Heynen, N. 2014. Urban political ecology I: The urban century. *Progress in Human Geography* 38 (4): 598–604.

Holifield, R. 2009. Actor-network theory as a critical approach to environmental justice: A case against synthesis with urban political ecology. *Antipode* 41: 637–658.

———. 2015. Environmental justice and political ecology. In *The Routledge handbook of political ecology*, ed. T. Perreault, G. Bridge, and J. McCarthy. London: Routledge.

Holifield, R., J. Chakraborty, and G. Walker. 2018. Introduction: The worlds of environmental justice. In *The Routledge handbook of environmental justice*, ed. R. Holifield, J. Chakraborty, and G. Walker. London: Routledge.

Honneth, A. 1995. *The struggle for recognition: The moral grammar of social conflicts*. Cambridge: MIT Press.

———. 2001. Recognition or redistribution? Changing perspectives on the moral order of society. *Theory, Culture, and Society* 18 (2–3): 43–55.

Horkheimer, M. 1970. *Traditionelle und kritische Theorie: Vier Aufsätze*. Frankfurt am Main: Fischer Bücherei.

Igoe, J. 2017. *The nature of spectacle: On images, money, and conserving capitalism*. Tucson: University of Arizona Press.

Isaac, J.C. 1987. *Power and Marxist theory. A realist view*. Ithaca: Cornell University Press.

Johnsen, K.I., and T.A. Benjaminsen. 2017. The art of governing and everyday resistance: 'Rationalization' of Sámi reindeer husbandry in Norway since the 1970s. *Acta Borealia* 34 (1): 1–25.

Kallis, G. 2017. *Defense of degrowth: Opinions and minifestos*. Nevele (Belgium): Uneven Earth Press.

———. 2019. *Limits: Why Malthus was wrong and why environmentalists should care*. Palo Alto: Stanford University Press.

Kelly, A.B. 2011. Conservation practice as primitive accumulation. *Journal of Peasant Studies* 38 (4): 683–701.

Köckler, H., S. Deguen, A. Ranzi, A. Melin, and G. Walker. 2018. Environmental justice in Western Europe. In *The Routledge handbook of environmental justice*, ed. R. Holifield, J. Chakraborty, and G. Walker. London: Routledge.

Larson, B. 2014. *Metaphors for environmental sustainability: Redefining our relationship with nature*. New Haven: Yale University Press.

Latour, B. 2005. *Reassembling the social: An introduction to actor-network theory*. Oxford: Oxford University Press.

Lave, R. 2015. Reassembling the structural: Political ecology and actor-network theory. In *The Routledge handbook of political ecology*, ed. T. Perreault, G. Bridge, and J. McCarthy, 213–223. London: Routledge.

Lawhon, M. 2013. Situated, networked environmentalism: A case for environmental theory from the south. *Geography Compass* 7 (2): 128–138.

Leach, M., and R. Mearns, eds. 1996. *The lie of the land: Challenging received wisdom on the African environment*. Oxford: James Currey.

Lecuyer, L., R.M. White, B. Schmook, V. Lemay, and S. Calmé. 2018. The construction of feelings of justice in environmental management: An empirical study of multiple biodiversity conflicts in Calakmul, Mexico. *Journal of Environmental Management* 213: 363–373.

Li, T.M. 2007. The will to improve. In *Governmentality, development, and the practice of politics*. Durham: Duke University Press.

Li, T.M. 2009. To make live or die? Rural dispossession and the protection of surplus populations. *Antipode* 41 (1): 66–93.

Lukes, S. 2005. *Power: A radical view*. 2nd ed. New York: Palgrave Macmillan.

Malthus, T.R. 1798 (1966). *First essay on population*. London: Macmillan.

Mansfield, B. 2010. Is fish health food or poison? Farmed fish and the material production of un/healthy nature. *Antipode* 43 (2): 413–434.

Mariki, S., H. Svarstad, and T.A. Benjaminsen. 2015. Elephants over the cliff: Explaining wildlife killings in Tanzania. *Land Use Policy* 44: 19–30.

Martin, A., S. McGuire, and S. Sullivan. 2013. Global environmental justice and biodiversity conservation. *Geographical Journal* 179: 122–131.

Martin, A., B. Coolsaet, E. Corbera, N.M. Dawson, J.A. Fraser, I. Lehmann, and I. Rodriguez. 2016. Justice and conservation: The need to incorporate recognition. *Biological Conservation* 197: 254–261.

Martinez-Alier, J. 2002. *The environmentalism of the poor: A study of ecological conflicts and valuation*. Cheltenham: Edward Elgar.

Martinez-Alier, J., I. Anguelovski, P. Bond, D. Del Bene, F. Demaria, J.-F. Gerber, L. Greyl, W. Haas, H. Healy, V. Marín-Burgos, G. Ojo, M. Porto, L. Rijnhout, B. Rodríguez-Labajos, J. Spangenberg, L. Temper, R. Warlenius, and I. Yánez. 2014. Between activism and science: Grassroots concepts for sustainability coined by Environmental Justice Organizations. *Journal of Political Ecology* 21: 19–60.

Martinez-Alier, J., L. Temper, D. Del Bene, and A. Scheidel. 2016. Is there a global environmental justice movement? *Journal of Peasant Studies* 43 (3): 731–755.

Marx, K. 1852. The Eighteenth Brumaire of Louis Napoleon. https://www.marxists.org/archive/marx/works/download/pdf/18th-Brumaire.pdf.

———. 1990. *Capital: A critique of political economy*. Vol. 1. New York: Penguin.

McCarthy, J. 2005. First World Political Ecology: Directions and Challenges. *Environment and Planning A*, 37: 953–958.

Meadows, D.H., D.L. Meadows, J. Randers, and W.W. Behrens III. 1972. *Limits to growth*. New York: Universe Books.

Meek, D., and T. Lloro-Bidart. 2017. Introduction: Synthesizing a political ecology of education. *Journal of Environmental Education* 48 (4): 213–225.

Mehta, L., J. Allouche, A. Nicol, and A. Walnycki. 2014. Global environmental justice and the right to water: The case of peri-urban Cochabamba and Delhi. *Geoforum* 54: 158–166.

Meillassoux, C. 1981. *Maiden, meal, and money: Capitalism and the domestic community*. Cambridge: Cambridge University Press.

Mingay, G.E. 1997. *Parliamentary enclosure in England*. London: Longman.

Neimark, B., J. Childs, A.J. Nightingale, C.J. Cavanagh, S. Sullivan, T.A. Benjaminsen, S. Batterbury, S. Koot, and W. Harcourt. 2019. Speaking power to 'post-truth': Critical political ecology and the new authoritarianism. *Annals of the American Association of Geographers* 109 (2): 613–623.

Netting, R. McC. 1977. *Cultural ecology*. Menlo Park: Cummings.

———. 1981. *Balancing on an alp: Ecological change and continuity in a Swiss mountain community*. Cambridge: Cambridge University Press.

———. 1993. *Smallholders, householders: Farm families and the ecology of intensive, sustainable agriculture*. Stanford: Stanford University Press.

Neumann, R.P. 2008. Probing the (in)compatibilities of social theory and policy relevance in Piers Blaikie's political ecology. *Geoforum* 39 (2): 708–715.

Paulson, S., and L. Gezon, eds. 2005. *Political ecology across spaces, scales and social groups*. New Brunswick: Rutgers University Press.

Peet, R., and M. Watts, eds. 1996. *Liberation ecologies. Environment, development, social movements*. London: Routledge.

Pellow, D.N. 2002. *Garbage wars. The struggle for environmental justice in Chicago*. Cambridge, MA and London, UK: The MIT Press.

———. 2016. Toward a critical environmental justice studies. Black lives matter as an environmental justice challenge. *Du Bois Review* 13 (2): 221–236.

———. 2018. *What is critical environmental justice?* Cambridge, UK and Medford, MA: Polity Press.

Pellow, D.N., and R.J. Brulle, eds. 2005. *Power, justice, and the environment: A critical appraisal of the environmental justice movement*. Boston: MIT Press.

Pulido, L. 1996. *Environmental and economic justice. Two Chicano struggles in the southwest*. Tucson: The University of Arizona Press.

Pulido, L., and J. De Lara. 2018. Reimaging 'justice' in environmental justice: Radical ecologies, decolonial thought, and the Black Radical Tradition. *Environment and Planning E: Nature and Space* 1 (1–2): 76–98.

Ranganathan, M., and C. Balazs. 2015. Water marginalization at the urban fringe: Environmental justice and urban political ecology across the north-south divide. *Urban Geography* 36 (3): 403–423.

Rawls, J. 1999. *A theory of justice*. Cambridge, MA: The Belknap Press of Harvard University Press.

Rigg, J. 2006. P. Blaikie 1942–. In *Fifty key thinkers on development*, ed. D. Simon. London: Routledge.

Robbins, P. 2020a. *Political ecology. A critical introduction*. 3rd ed. Oxford: Wiley Blackwell.

———. 2020b. Is less more … or is more less? Scaling the political ecologies of the future. *Political Geography* 76: 102018.

Rocheleau, D. 2008. Political ecology in the key of policy: From chains of explanation to webs of relation. *Geoforum* 39 (2): 716–727.

Roe, E. 1991. Development narratives, or making the best of blueprint development. *World Development* 19 (4): 287–300.

———. 1999. *Except-Africa. Remaking development, rethinking power*. New Brunswick: Transaction Publishers.

Ross, E.B. 1998. *The Malthus factor: Poverty, politics and population in capitalist development*. London: Zed Books.

Said, E. 1978. *Orientalism. Western conceptions of the orient*. London: Penguin.

Salazar Parrenas, J. 2018. *Decolonizing extinction. The work of care in orangutan rehabilitation*. Durham: Duke University Press.

Sayre, N.F. 2017. *The politics of scale. A history of rangeland science*. Chicago: University of Chicago Press.

Schlosberg, D. 2004. Reconceiving environmental justice: Global movements and political theories. *Environmental Politics* 13 (3): 517–540.

———. 2007. *Defining environmental justice: Theories, movements, and nature*. Oxford: Oxford University Press.

Schlosberg, D., L. Rickards, and J. Byrne. 2018. Environmental justice and attachment to place: Australian cases. In *The Routledge handbook of environmental justice*, ed. R. Holifield, J. Chakraborty, and G. Walker. London: Routledge.

Schroeder, R., K. St. Martin, and K.E. Albert. 2006. Political ecology in North America: Discovering the third world within? *Geoforum* 37: 163–168.

Schroeder, R., K. St. Martin, B. Wilson, and D. Sen. 2008. Third world environmental justice. *Society and Natural Resources* 21 (7): 547–555.

Schulz, K.A. 2017. Decolonizing political ecology: Ontology, technology and 'critical' enchantment. *Journal of Political Ecology* 24: 125–143.

Scott, J.C. 1976. *The moral economy of the peasants: Rebellion and subsistence in Southeast Asia*. New Haven: Yale University Press.

———. 1985. *The weapons of the weak: Everyday forms of peasant resistance*. New Haven: Yale University Press.

———. 1990. *Domination and the arts of resistance: Hidden transcripts*. New Haven: Yale University Press.

Shaw, A. 2018. Environmental justice for a changing Arctic and its original peoples. In *The Routledge handbook of environmental justice*, ed. R. Holifield, J. Chakraborty, and G. Walker. London: Routledge.

Sikor, T., ed. 2013. *The justices and injustices of ecosystem services*. London: Earthscan.

Sikor, T., and P. Newell. 2014. Globalizing environmental justice? *Geoforum* 54: 151–157.

Simon, D. 2008. Political ecology and development: Intersections, explorations and challenges arising from the work of Piers Blaikie. *Geoforum* 39 (2): 698–707.

Steger, T., R. Filcák, and K. Harper. 2018. Environmental justice in Central and Eastern Europe: Mobilization, stagnation and detraction. In *The Routledge handbook of environmental justice*, ed. R. Holifield, J. Chakraborty, and G. Walker. London: Routledge.

Stern, M., S. Hellberg, and S. Hansson. 2015. Studying the agency of being governed: An introduction. In *Studying the agency of being governed*, ed. S. Hansson, S. Hellberg, and M. Stern. New York: Routledge.

Stott, P., and S. Sullivan, eds. 2000. *Political ecology: Science, myth and power*. London: Arnold.

Svarstad, H. 2009. Narrativitetens sosiologi. *Sosiologi i dag* 39 (4): 29–56.

Svarstad, H., and T.A. Benjaminsen. 2020. Reading radical environmental justice through a political ecology lens. *Geoforum* 108: 1–11.

Swyngedouw, E., and N.C. Heynen. 2003. Urban political ecology, justice and the politics of scale. *Antipode* 35 (5): 898–918.

Temper, L. 2019. Blocking pipelines, unsettling environmental justice: From rights of nature to responsibility to territory. *Local Environment* 24 (2): 94–112.

Turner, M. 1993. Overstocking the range: A critical analysis of the environmental science of Sahelian pastoralism. *Economic Geography* 69 (4): 402–421.

———. 1998. The interaction of grazing history with rainfall and its influence on annual rangeland dynamics in the Sahel. In *Nature's geography. New lessons for conservation in developing countries*, ed. K. Zimmerer and K.R. Young. Madison: University of Wisconsin Press.

Valdivia, G. 2015. Eco-governmentality. In *The Routledge handbook of political ecology*, ed. T. Perreault, G. Bridge, and J. McCarthy. London and New York: Routledge.

van der Ploeg, J.D. 2013. *Peasants and the art of farming: A Chayanovian manifesto*. Bourton on Dunsmore: Fernwood Publishing.

Vayda, A. 1983. Progressive contextualization: Methods for research in human ecology. *Human Ecology* 11 (3): 265–281.

Vayda, A.P., and B. Walters. 1999. Against political ecology. *Human Ecology* 27 (1): 167–179.

Vermeylen, S. 2019. Environmental justice and epistemic violence. *Local Environment* 24 (2): 89–93.

Wainwright, J. 2008. *Decolonizing development: Colonial power and the Maya*. Oxford: Blackwell.

Walker, P.A. 2005. Political ecology: Where is the ecology? *Progress in Human Geography* 29 (1): 73–82.

Walker, G. 2012. *Environmental justice: Concepts, evidence and politics*. New York: Routledge.

Watts, M. 1983. *Silent violence: Food, famine and peasantry in northern Nigeria*. Berkeley: University of California Press.

———. 1997. Classics in human geography revisited. *Progress in Human Geography* 21 (1): 75–80.

World Commission on Environment and Development (WCED). 1987. *Our common future*. Oxford: Oxford University Press.

Young, I. 1990. *Justice and the politics of difference*. Princeton: Princeton University Press.

Zimmerer, K.S. 2013. Cultural ecology and human ecology. *Oxford Bibliographies*. https://doi.org/10.1093/OBO/9780199874002-0041.

Discourses and Narratives on Environment and Development: The Example of Bioprospecting

Contents

T. A. Benjaminsen, H. Svarstad, *Political Ecology*,
https://doi.org/10.1007/978-3-030-56036-2_3

Trailer

What is bioprospecting? Let us start with the Hollywood version. Deep in the rainforest of Brazil an old traditional healer lives, who has a herbal cure for cancer. *Medicine Man* is a film classic from 1992 in which the legendary actor Sean Connery plays Dr. Robert Campbell, a scientist who has learned about that herbal cure. He is soon joined by a woman who also is a scientist, Dr. Rae Crane, played by Lorainne Bracco. The two not only have a romance going, but they take part in a global activity. This activity is what we call bioprospecting.

When conducting bioprospecting, scientists connected to the pharmaceutical industry travel the globe to collect biological samples. Like Dr. Campbell in the film, real bioprospectors sometimes acquire knowledge from traditional healers on organisms that they use to prepare medicines to treat various diseases. The natural medicines then provide the starting points for further laboratory analyses that aim to develop modern medicines.

In the film, modern development approaches in terms of giant road construction machines, and more and more of the forest is burned down, and the natural source of the cancer medicine is threatened. The two bioprospectors involve themselves in a dramatic battle in order to ensure the survival of the local indigenous people as well as of the cancer medicine.

But in the film, we gradually learn that the work of Dr. Campbell has not always had positive consequences. A few years earlier, he visited a neighbouring tribe in his search for traditional knowledge to forward to the American pharmaceutical company that he worked for. He found that this indigenous group had a pain reliever. The pharmaceutical company set up a large laboratory with many research workers in the forest and developed a medicine that became a financial success. However, for the indigenous group the consequences were fatal. From the laboratory staff they contracted influenza, against which they had no resistance. All died. (◘ Photo 3.1)

Overview

In this chapter, we tell our version of the story of bioprospecting that in recent decades has constituted an important and contested topic within the field of environment and development. We also use bioprospecting as an example to illustrate how elements may be studied and connected within the framework of political ecology. First, we discuss the political economy of bioprospecting. Then we present two sets of discourses and narratives that have dominated ways of telling about the economic activity and its consequences. We continue by presenting a case study from Tanzania of what bioprospecting there entailed for involved people, and we compare these findings with each of the two leading discourses and related narratives about bioprospecting. Moreover, we tell the story about a new medicine that began on a mountain road in Norway and ended with the most-selling medicine of the largest pharmaceutical company in the world.

3

Photo 3.1 In the film *Medicine Man*, the relationship between the bioprospector Dr. Campbell and local people is presented as harmonious as well as problematic. (Credit: CINERGI/COLUMBIA/TRI-STAR//O'NEILL, TERRY/Album)

3.1 The Political Economy of Bioprospecting

The film *Medicine Man* was inspired by a new wave of bioprospecting that took off in the early 1990s as a strategy for developing medicines. The search for biological material for all kinds of human uses is probably as old as humanity and an aspect of social development in all societies. The great European voyages of discovery aimed not just to explore the globe and conquer the world, but also to acquire knowledge of natural resources that could be used to produce food, medicines, building materials and other essentials. Botanical gardens and zoos systematized specimens collected, which created a basis for economic exploitation (Foucault 1970; Parry 2000). The Kenyan social scientist Calestous Juma (1989) points out that such collections, along with the development of technologies and institutions using them, comprised an essential though overlooked role in the economic growth of the colonial powers.

In other words, the pharmaceutical industry's focus on bioprospecting from the 1990s was a continuation of an old activity. For decades, pharmaceutical industry had focused on more chemically oriented strategies, but new technology again made it interesting and potentially profitable to turn to nature as a starting point for developing modern medicines. Moreover, new property rights associated with biological building blocks made bioprospecting more attractive. Therefore, in the 1990s, all major global pharmaceutical companies became involved in bioprospecting, either by themselves, or jointly with smaller companies specializing in bioprospecting, or together with research institutes.

Bioprospecting is an economic activity with a clear North-South dimension. Companies that use bioprospecting to develop new products are mainly found in wealthy countries in the Global North. At the same time, many tropical countries in the Global South possess a large biological diversity, with many different species of plants and animals, and much genetic variation also within species. This has made tropical countries particularly attractive for bioprospectors.

The activity involves various actors, including agents who travel from the Global North to the South to collect samples of biodiversity, and people in the South who help the agents collect plant samples and sometimes teach them how the plants are used in traditional medicine. These actors meet face-to-face. At the same time, places far from each other become linked together, such as collection areas in Africa and South America, and research institutes and pharmaceutical companies in Europe and North America.

From the early 1990s on, bioprospecting also came into focus in international negotiations and conferences on biodiversity. These were first of all negotiations and follow-up of the Convention on Biological Diversity (CBD). Many leaders in developing countries were optimistic, because they saw that their abundance of biodiversity had become an asset in demand as a raw material for companies in developed countries. This gave rise to thoughts of biodiversity as a new green gold that could provide income for developing countries. At the same time, the notion of neo-colonial exploitation linked to this activity arose.

At the summit for environment and development in Rio de Janeiro in 1992, as well as during previous negotiations and at the Conferences of the Parties to the CBD every other year afterwards, there have been numerous discussions about possible consequences and possibilities to regulate bioprospecting. In 2007, most countries still lacked laws and procedures necessary for profiting from bioprospecting, although some 58 countries were on the verge of or had implemented relevant legislation (CBD 2007; Wynberg and Laird 2009).

In 2014, the Nagoya Protocol came into force. This protocol specifies conditions for bioprospecting and rights of source countries of biodiversity to take part in benefits from the access and use of genetic resources.

After three decades, in 2020 bioprospecting can be concluded to have been a wave that did not lead to the growth of activities and income that some hoped for. An important reason for the decline was that critique and protests against the activity made companies prefer other alternatives. At the same time, new technology has again made chemical methods attractive for pharmaceutical companies. Nevertheless, in this chemical approach, companies often still apply knowledge that they have already collected through bioprospecting, when they elaborate new products based on imitations of nature (Neimark 2017).

We see the bioprospecting wave as an example of the rise, fall and change of an industrial activity with expectations of income and development not only for companies in the Global North, but also for the Global South. Recently, newer waves of such activities have emerged, and some of them with new promises for the South. In ▸ Chap. 6, we discuss a case of the use of tropical forests as carbon sinks to mitigate climate change with promises of payments from the Global North. Looking at the history of such new trends, it is interesting to see how some ways of understanding an issue—or discourses—evolve in parallel to industrial developments.

The Discourse Concept

We define a *discourse* as a socially shared perspective of a topic. A discourse may be likened to a lens through which a topic is viewed. Each discourse is based on assumptions, arguments and claims about major aspects of an issue. People create discourses through their written and spoken statements. Within each discourse, there are statements with similarities of content and often also use of the same means of expression, such as particular metaphors and ways of relating concrete examples. Discourses comprise structures that are created, maintained and changed by people and may be produced by smaller or larger groups of people at the local, national or even the global level.

Some discourses create essential frameworks for interpreting particular cases. Such discourses influence policies, laws and practices, and we call them *leading discourses*. Actors who are instrumental in creating, altering and communicating such discourses, exercise discursive power.

In a discourse analysis, we examine statements in order to identify and describe discourses. When we use discourse analysis in our own research, we always combine it with other types of research. Often this involves examinations of central claims of leading discourses on an issue. For example, an analysis of discourses on bioprospecting can be followed by examining key assertions of how bioprospecting is conducted in practice, and of the consequences the activities have for various groups of people.

Michel Foucault provided seminal contributions to the notion of discourse, from which a broad variety of discourse perspectives and analyses have evolved in the social sciences and the humanities. Through historical studies of topics, such as of the treatment of insanity and of punishment and incarceration, Foucault (1979, 1988) showed how government practices in various periods have been linked to discourses on what is regarded to be meaningful, true and accepted forms of statements and actions.

Antonio Gramsci provided similar ideas about ways that rulers may get their perspectives adopted by the citizens (▸ Chap. 1). In situations with limited openness for exchanges of ideas, many topics are easily dominated by only one discourse. In democratic countries, we might expect different opinions and the production of more than one discourse about important topics (but that is not always the case, as we show in an example in ▸ Chap. 6).

On issues of governance of land and natural resources, we often find that two or a few more discourses have been important in recent times. This chapter shows the example of bioprospecting, which has been dominated by two opposing discourses since the beginning of the 1990s.

It can be useful to locate the discourse concept relative to other key concepts of human opinion and thought. Original views on a theme put forth by an individual will lie outside established discourses. Discourses are *social* constructions, in other words they are produced by a number of people together. That said, the views of an individual may contribute substantially in establishing, producing, or transforming discourses.

Ideology is a term often used on the main ideas about how society should be organized. Liberalism, socialism and nationalism are examples of such ideas. Ideologies contrast the concept we here introduce about discourses as perspectives on more limited themes.

Culture may have much in common with the discourse concept. Both concern meaning constructions shared by many individuals, but culture usually is used in a broader sense and encompasses deeper structures than discourses. The culture of a society includes fundamental ways of perceiving, living and behaving in society.

It is worthwhile emphasizing that discourses constitute important *social structures*. Hence, they contribute both to constraining and enabling actions (► Chap. 1).

Paradigm is a label often used interchangeably with discourse. In such cases, the discourse is scientific, that is, it is a scientific discipline or approach within a discipline. However, many discourses are produced outside the world of science, and actors other than scientists are key contributors.

Like discourses, *theories* contain visions of reality. Usually, only scientists have the authority to establish theories. Discourses distinguished by actors other than scientists often tend to be more influential on practices, and an example is the governance of natural resources. Therefore, we argue that it is important for students not only to study topics in comparison with theories, as it is usually done, but also to compare practices with the leading discourses.

In this book, we use a discourse concept often drawn on in the social sciences. It is worthwhile to recognize that this differs from the everyday use of the term discourse as a synonym for discussion. Actually, discourse is the opposite of discussion. A discussion involves differences of opinion, whereas the concept of discourse that we use instead implies a pattern of similarities in statements.

Moreover, there is a range of linguistic perspectives on discourse, and they are much more engaged with the ways in which language is used. Linguists often see conversation to constitute a discourse, and an analysis of discourse may focus on how language is used in producing opinion. Social scientists also often apply elements of linguistic discourse analyses.

But the purpose of discourse analysis in the social sciences is not to produce knowledge about language as such. The purpose is rather to chart to what extent and how discourses structure our views on particular topics, and to examine whether there are particular groups of actors who exercise power by contributing substantially to this structuration.

In environmental studies, Maarten Hajer (1995) and John Dryzek (1997) introduced discourse analyses that have later been applied and further elaborated by other scholars in the field.

3

Within a topic, we may identify and analyse which discourses are produced. We call a collection of discourses on a topic a *discourse order* (Fairclouch 1995). Furthermore, we find it useful to put a particular emphasis on *leading discourses*. These discourses have consequences in terms of political and administrative decisions, and thereby, for instance, for the governance of natural resources.

Moreover, we can consider a discourse to be *hegemonic* if there are no other leading discourses in the same field, so that this discourse dominates policy-making and practice. Actors who contribute to leading or hegemonic discourses exercise *discursive power*.

In the following, we present two discourses that during a large period have been leading on the topic of bioprospecting. Both these discourses have influenced elements of politics and practices without being dominating. To identify these discourses, empirical data have been collected through participant observation at international meetings and conferences, as well as by examination of written sources.

3.2 The Bioprospecting Win-Win Discourse

There is a bioprospecting win-win discourse entailing a message that bioprospecting is an activity which is beneficial for all actors involved. Companies earn money from the sale of new medicines that they develop. The medicines benefit patients, and there has been a focus on remedies to combating cancer and HIV/AIDS. Furthermore, from the win-win discourse it is argued that given the introduction of specific regulations, bioprospectors will pay part of the incomes accrued from new medicines to the countries in which the relevant bioprospecting has taken place. This often means tropical countries in the Global South that have a great biological diversity and extensive traditional knowledge about medicinal effects of plants. Besides bioprospectors themselves, conservation organizations have played a key role in establishing this discourse. Conservationists have argued that revenues from bioprospecting should in part go both to finance conservation and to compensate the local collaborators of the bioprospectors. In this way, people in areas of abundant biodiversity will see that preserving diversity leads to economic benefits.

From the early 1990s, the win-win discourse integrated elements of discourse production by four groups of actors. The first group consists of scientists specialized in biological and chemical studies of natural products, including traditional medicines. The second group includes scientists concerned with conservation biology, a field of biology concentrated on methods for protecting nature, while the third includes conservationists in organizations such as the World Wide Fund for Nature (WWF), the International Union for Conservation of Nature and Natural Resources (IUCN) and the World Resources Institute (WRI), and the fourth consists of pharmaceutical companies.

Scientists are interested in greater focus on and financing of their specializations. Likewise, conservationists are concerned with promoting and obtaining funding for conservation, and pharmaceutical companies developing new medicines based on bioprospecting have an interest in getting access to biological diversity and to ensure social acceptance of their activities.

We have seen how the win-win discourse has involved each of the four groups arguing for the importance of bioprospecting without reference to their own self-interests. Instead, they argue that they are concerned with a win-win activity for a good cause, such as conservation, developing new medicines and providing income for poor countries and especially for rural poor in these countries.

Cornell University professor Thomas Eisner (1929–2011) was one of the earliest contributors to the win-win discourse on bioprospecting. He wrote enthusiastically about what he called chemical prospecting, which he believed should be intensified. He expressed concern for the extinction of species and thereby the loss of chemical components of great value in developing new products, such as medicines. Therefore, Eisner proposed that laboratories be set up in countries in the Global South to chart the components and at the same time support conservation programmes (Eisner 1989, 1991).

Furthermore, the Biodiversity Convention has played a key role for the elaboration of the win-win discourse on bioprospecting. The three goals of this convention are conservation of biodiversity, sustainable use of its components and the fair and equitable sharing of benefits from the use of genetic resources (United Nations 1992, article 1). The four groups of actors behind the win-win discourse have pointed out that bioprospecting can help attain these three goals. A key book on bioprospecting from World Resources Institute (WRI) states that:

> Done well, biodiversity prospecting can contribute greatly to environmentally sound development and return benefits to the custodians of genetic resources – the national public at large, the staff of conservation units, the farmers, the forest dwellers, and the indigenous people who maintain or tolerate the resources involved (Reid et al. 1993: 2).

Contributors to the win-win discourse has often pointed to the importance of establishing institutional frameworks in the countries in which bioprospecting takes place. Thus, they have emphasised on putting laws and administrative apparatus in place in order to ensure that bioprospectors in one way or another give something in return to the source countries of biodiversity as well as to the local participants involved.

3.3 The Biopiracy Discourse

Solidarity organizations for marginalized people in the Global South have played a key role in initiating and producing the biopiracy discourse. The most important of these organizations are the ETC Group from North America (formerly called the Rural Advancement Foundation International – RAFI), the Genetic Resources Action International (GRAIN) from Europe and the Third World Network from Asia. In this discourse, bioprospecting is regarded as exploitation carried out by industrial companies and their home countries in the North. The losers are small-scale farmers, indigenous peoples and traditional healers who are duped into giving away valuable natural resources and associated knowledge with no more than chicken feed as payment.

Biopiracy refers to the monopolization of genetic resources such as seeds and genes taken from peoples or farming communities that have nurtured those resources through generations. It also refers to the theft of traditional knowledge from those cultures (Coalition Against Biopiracy 2008). (◻ Fig. 3.1)

The biopiracy discourse evolved in response to the increased focus on bioprospecting of the 1990s. Pat Roy Mooney, the Canadian director of the ETC Group and co-founder of its predecessor, RAFI, has long been concerned with questions about plant breeders' rights regarding agricultural crops. According to Mooney (2000), the Group coined the term biopiracy to counterbalance the favourable presentations of bioprospecting by Reid et al. (1993). In a pamphlet published by RAFI, they calculated a large loss for tropical countries because countries in the North use seeds and medicines from the South without paying for their true worth (RAFI 1994). Biopiracy is depicted as a global pandemic with the contention that '... there are few places on earth where rural people are not facing biopirates who aim to extract their knowledge and resources' (RAFI 1995: 1). In a press release from Inter Press Service, biopiracy was defined as 'the stealing of plants for commercial purposes' (Portillo 1999).

The win-win discourse has an institutional focus on the establishment of legislation in developing countries. This is in accordance with an aim of the Biodiversity Convention of ensuring gains for source countries for bioprospecting. On the other hand, the biopiracy discourse has an institutional focus on patenting as a means of exploitation (Svarstad 2000, 2003, 2005).

Biopatents comprise a company's property rights to genetic material that have resulted from new technical alterations. A company takes out a patent on genetic material in order to get exclusive rights to it in one or more countries and thereby

◻ **Fig. 3.1** At several international conferences, the Coalition Against Biopiracy has arranged ceremonies where 'winners' are announced of Captain Hook Awards for Biopiracy. This poster is from the Conference of the Parties to the Convention on Biological Diversity (CBD) in Cacun, Mexico, in 2016. (SynBioWatch 2016)

clude competition from other companies in those countries. A patent holder may choose to sell licenses that require payment for use of the genetic material in production, such as of a particular medicine. Patent rights are valid in limited periods of time, and usually about 20 years.

In GRAIN's Seedling magazine, Janet Bell (1997) wrote that the Bioconvention's goal of equitable sharing of benefits gained from biological diversity is 'a difficult goal if the bioprospector can slash intellectual property rights over whatever he/she finds' (Bell 1997: 2).

The Indian activist Vandana Shiva has for many years been a leading opponent of bioprospecting and its associated patenting (1997, 2007). She has characterized the introduction of intellectual property rights to forms of life as a means to legitimate biopiracy, and she contends that patents 'protect this piracy of the wealth of non-western peoples as a right of Western powers' (Shiva 1997: 5).

The British geographer Noel Castree shares our description of the polarized debate on bioprospecting as the production of two rival discourses (Castree 2003). Some academic contributors to the biopiracy discourse have described the economic activity of bioprospecting as a form of neoliberalism, because businesses play a prominent role, markets for genetic materials are established, and patenting being a form of privatization (see for instance Brush 1999; Martinez-Alier 1996). Castree criticizes such contributors for not being constructive in their critiques in terms of not proposing alternative courses of action.

Discursive Narratives

In each of the two discourses on bioprospecting, narratives play a significant role. Before presenting examples, we will explain what this implies theoretically. We define a *narrative* as a story that has a *storyline* and with the involvement of one or more actors in a *gallery of actors*. We are also interested in the *narrative producers,* and these are actors who contribute to the creation, reproduction and modifications of narratives.

Some narratives are pure fiction, as in novels, fairy tales and films (such as *Medicine Man* and *Avatar*). Other narratives contain claims about reality. The sociologist Anna Johansson points out that to narrate is a universal human activity, and that narratives can be regarded as a basic form of culture (Johansson 2005).

The focus on the storyline (or plot) of a narrative may be traced back to the Greek philosopher Aristotle of the fourth century BC. He defined a narrative as a story with a beginning, a middle and an end (Aristotle 1996). The storyline is not just a sequence of events, but it also contains claims of causal relationships (Elliott 2005; Johansson 2005).

The actor gallery of a narrative may sometimes have similarities with classical fiction characters, such as heroes, villains and victims (Svarstad 2000; Elliott 2005).

A *discursive narrative* is a story about a specific case of an issue, where the case is interpreted through the perspectives of a discourse. While drawn from a discourse, a discursive narrative also contributes to the production of the discourse.

In other words, a discourse may create foundations for interpretation of concrete cases, and at the same time, they are used as proof of perspectives and claims of the discourse.

The production of discourses as well as of discursive narratives provides examples of the agency-structure relation we presented in ▶ Chap. 1. Some actors may consciously contribute to the production of discursive narratives in order to communicate their messages. This is, in other words, an attempt to exert discursive power (also presented in ▶ Chap. 1). Others may learn about a case of an issue without realizing that this is based on specific claims, interpretations and agendas.

There are other narrative concepts with similarities and differences compared to that presented here. Emery Roe has played a seminal role in focusing on narratives in analyses relevant to natural resources and development (Roe 1991, 1995, 1999). We believe his focus is important, even though his conceptual apparatus differs from ours.

Roe's main narrative concept is similar to what we may call a storyline of a discourse. This provides a central element of a discourse and relates the course of action that may happen if certain premises are met. Thus, *development narratives*, as Roe calls them, describe what will happen from a given starting point. One such development narrative is 'Everything works ... except in Africa' (Roe 1999: 2). Roe rightfully criticizes this and other development narratives for having a weak empirical basis.

Our concept of discursive narrative differs from Roe's *narrative* by providing claims about what has happened or is happening in specific cases of an issue, and with a clearer actor perspective. Roe uses his narrative concept about more general claims of development directions given a specific start.

Furthermore, in his book on 'the postmodern condition', French philosopher Jean-François Lyotard argued that society is on its way to a *postmodern* state in which what he called 'grand narratives' or 'metanarratives' are on the way out. Instead, 'small narratives' that are negotiated locally will more and more take over (Lyotard 1997). Lyotard's concept of grand narratives refers to something more wide-ranging than discursive narratives or discourses presented in this book. Lyotard referred rather to general ideologies, such as Christianity, Islam, communism and capitalism.

We do not, however, agree that society is evolving as Lyotard argued. Recent discourses on environmental governance for instance can to a great extent be associated with wider ideologies. Moreover, we see that they are widespread as an aspect of the present cultural globalization. Many of these discourses embrace neoliberalism and economic globalization, while others tend to be fundamentally critical to the these. Lyotard's argument of the greater role of small narratives stands in contrast to observations that global actors often strongly influence practices at the local level with their interpretation of situations through discursive narratives. These narratives usually deal with local conditions, but draw from discourses with a global coverage.

3.4 Win-Win Narratives on Bioprospecting

The win-win discourse on bioprospecting contains discursive narratives that describe examples of bioprospecting in a positive manner. A key example consists of variations over a positive story of the cooperation between the Costa Rican research institute INBio and the giant American pharmaceutical company Merck. In 1991, the two parties made a cooperation agreement where INBio was paid 1.1 million US dollars for collecting and preparing biological specimens for Merck. According to the agreement, INBio would also receive a percentage of revenues generated from sales if new medicines resulted from the cooperation. The agreement has been emphasized as a show case of bioprospecting and how this activity can contribute to conservation, because INBio gave part of its income to the conservation authorities in Costa Rica (Mateo 2000; Reid et al. 1993). At a meeting in Geneva in 1991, one of us observed how the agreement was presented directly to the negotiators involved in the concluding work of the Biodiversity Convention.

In 1990, Shaman Pharmaceuticals, an American company, was founded to work exclusively on bioprospecting. The company worked in local communities in Latin America, Africa and Southeast Asia, until it in 2002 went bankrupt. Shaman based all its collections of plant samples on information acquired from traditional healers and other local actors with expertise on medicinal effects of plants. The samples were sent to Shaman's laboratory in California where chemists and pharmacists investigated effects and side effects of the biochemicals. Shaman and an associated charity organization, Healing Forest Conservancy, presented several win-win narratives based on their experiences in bioprospecting. While INBio has emphasized conservation as a key aim for sharing the benefits of bioprospecting, Shaman emphasized that the receivers of such benefits should be 'indigenous, tribal or native' people (Moran 1997: 246). The company has presented win-win narratives of their bioprospecting in countries such as Ecuador (King 1994), Belize (Chinnock et al. 1997), Nigeria (Carlson et al. 1997) and Tanzania (Richter et al. undated).

One of Shaman's narratives concerns medicinal effects of the resin of species of the dragon blood tree (*Croton*) of South America. For a long time, the company worked to develop two anti-viral medicines, and described a successful cooperation in this case with an umbrella organization for indigenous peoples in South America (King 1994). Shaman developed medicines based on the dragon blood tree that were promising for treating widespread diseases and consequently could have a large potential market (Shaman Pharmaceuticals 1997a, b). Commercial production would require a large amount of raw material, and Shaman wanted indigenous people in forests with dragon blood trees to supply the raw material and thereby contribute to their livelihoods. The company also emphasized their concern for such production to be ecologically sustainable.

3.5 Biopiracy Narratives

Many biopiracy narratives have been produced within the biopiracy discourse. The individual narratives focus on various stages of bioprospecting, but all provide concrete examples of how bioprospecting and associated patenting is tantamount to exploitation. In some cases, bioprospecting is the starting point for both win-win and biopiracy narratives. As presented above, the agreement between INBio and Merck has been presented as a win-win. According to a biopiracy narrative of this case, the agreement implies that a multinational company, Merck, gains access to Costa Rica's genetic resources, while the country itself gains little and the indigenous peoples get nothing, even though they are entitled to compensation (Kloppenburg and Rodriguez 1992; Mooney 2000).

Shaman Pharmaceutical's win-win narratives have also been met with biopiracy narratives that we might regard as *counternarratives*, to use a term from Roe (1991, 1995). According to these narratives, Shaman underpaid traditional knowledge. Moreover, involved patents and Shaman's cooperation with major pharmaceutical companies have been spotlighted as elements of exploitation of peoples in the Global South.

For example, Shaman's win-win narrative on the dragon blood tree has been met by a counternarrative. Viki Reyes of GRAIN felt that in this case, the payment for traditional knowledge was problematic. In a series of pamphlets entitled *BioPirates*, RAFI and Cultural Survival Canada included Shaman's bioprospecting for the dragon blood tree arguing that Shaman acquired millions of dollars of venture capital in the USA, while the Quichua people of Jatún Molino in Ecuador received less than three thousand dollars from Shaman, much of which was pay for work and services (RAFI and Cultural Survival Canada 1997; Reyes 1996).

This biopiracy narrative also focused on Shaman's patents. According to the narrative, the company claimed an exclusive monopoly on a biochemical, despite its use being commonplace knowledge among several indigenous peoples of the Amazon area, such that it should not have been patentable (Reyes 1996).

Bioprospectors have presented many win-win narratives in scientific journals in the fields in which the bioprospectors themselves work. On the other hand, biopiracy narratives have often appeared in newsletters and on websites of the key organizations involved in the biopiracy discourse.

The actor gallery of win-win narratives consists of actors who contribute to the conduct of an activity with beneficial consequences. In contrast, the actor gallery of biopiracy narratives consists of actors who either are exploiters or have been exploited.

Elliott (2005) distinguishes between two basic types of narrative storylines. A *progressive* narrative relates progress, achievement and success, while a *regressive* narrative relates deterioration or recession. The win-win and biopiracy narratives correspond to these two types.

We have now provided examples of a group of win-win narratives on bioprospecting as well as a group of biopiracy narratives. Within each group, we have seen that there are similarities in the structures of the narratives. We have, in other words, found two *metanarratives*.

3.6 Back to Hollywood

If we go back to the *Medicine Man*, we can see that the film contains a main story that to a large extent is in accordance with the metanarrative of the win-win discourse. At the same time, the film also has a subplot on biopiracy. In the main plot, the bioprospectors work for two worthy causes. They strive to contribute to finding a medicine that will benefit cancer patients round the world. They also act as conservationists who try to prevent destruction of the rainforest in order to develop a cancer medicine as well as sustain the livelihoods of the indigenous group in the area. This group and their old medicine man are depicted as possessors of knowledge on biological resources of value to the entire world.

The details that gradually are revealed about Dr. Campbell's previous quest for medicine in a neighbouring tribe are in accordance with the metanarrative on biopiracy. The indigenous people there had a medicine that the American pharmaceutical company took advantage of and made a large profit from. For the local tribe the outcome was as adverse as it could be, as contact with the bioprospectors resulted in their contracting a disease from which all died.

It is interesting to see how the Hollywood movie in this way managed to employ principal aspects of both the two leading discourses on bioprospecting.

Investigating Reality Claims of Discourses and Narratives

How shall we deal with reality? We could put it in quotation marks and otherwise leave it to others to investigate. Regrettably, this is what some scholars engaged in studies of social constructions do. In political ecology, however, this should not be an alternative. One of the central aspects of this approach is to link social constructivist and realist knowledge (see the fifth synthesis in ► Chap. 1). We believe that it is particularly important to compare the social constructions on a topic that powerful actors produce with our own investigations of the same topic. This may involve case studies of actual practices, such as how bioprospecting takes place and with what consequences, or how conservation is carried out. It may also involve studies of the views among local communities, or investigations of particular aspects of natural conditions.

In our own research, we have found that both the win-win and the biopiracy narratives on bioprospecting have basic weaknesses. The producers of win-win narratives usually possess much knowledge of the various cases they present, because they have often been involved in the bioprospecting themselves. However, they do have clear interests in emphasizing positive aspects. Consequently, descriptions of harmonious cooperation and just distribution of benefits have often been exaggerated. On the other hand, many biopiracy narratives have been produced by actors with inadequate knowledge of the cases they describe. Despite these weaknesses, both win-win narratives and biopiracy narratives have through several decades been important for the production of the polarized couple of leading discourses on bioprospecting. This implies that the narratives have played important roles for decisions and structures about bioprospecting. We argue that it is important that researchers and students critically examine the empirical basis of leading discourses and thereby investigate the exercise of discursive power.

3.7 Tracking American Gene Hunters in Tanzania

In 1995, Shaman Pharmaceuticals organized a collection expedition to Tanzania. When one of us considered cases of bioprospecting to study, one of the company directors recommended this particular case. He was kind enough to send us the manuscript of an article about the expedition and its activities (Richter et al. undated). The manuscript contained a clear win-win narrative on bioprospecting. We took the director at his word, and in 1998, we followed the expedition's 'jeep tracks' in an area in the Kilimanjaro region. In our field work, we interviewed traditional healers and traditional midwives who had taken part in Shaman's bioprospecting and compared our finds with Shaman's account (Svarstad 2000, 2003, 2005).

During the collection, the expedition team had sought out people they assumed to have used their own plant remedies to treat patients, and they were asked if they used any such remedies to treat three particular diseases. Those who answered 'yes' were asked for further details on the plants involved, and finally the bioprospectors collected plant samples to send to Shaman's laboratory in California for analysis.

Shaman's narrative presented a successful and harmonious cooperation between external and local participants as equal partners. According to the manuscript, the company created economic equality and reciprocity. It donated a total of 6500 US dollars to local causes. In the area of our field work, the donation went to building materials for a centre for the district association of traditional healers and traditional midwives. In another area, the donation funded equipment for clinics and schools in the five villages involved. In addition, each of the local participants was paid five dollars a day. Furthermore, Shaman donated 14,000 dollars to the governmental Institute of Traditional Medicine, the Tanzanian research institute that had been involved.

The company also presented their intentions to donate funds to all its collaborating partners in various countries if their bioprospecting would result in income from sales of new products. Those donations were never made, as Shaman a few years later went bankrupt. Some of the other donations must be regarded as immediate payments for purchase of services.

On fieldwork in 1998, we found that the donation to the centre for traditional healers and traditional midwives had been made. However, the payment was insufficient to finish the centre, and the district association failed to find other donors. A small brick house was built halfway and then let decay. We interviewed many of the traditional healers in the area to learn about their opinions of the bioprospecting by Shaman Pharmaceuticals. Most of the interviewees provided ambivalent thoughts, although a few were very positive and a few very negative. The ambivalent healers wanted more bioprospecting, even though they were not content with the experiences so far.

We found four explanations for the ambivalence. First, economic benefits played a role, but in contrast to both of the leading discourses on bioprospecting, economic payment was of less importance than other matters.

Second, the traditional healers expressed eagerness to cooperate with bioprospectors as a way to access knowledge from modern science. Some healers regarded themselves as modern and scientific, whilst others put more emphasize on traditional aspects of their herbal medicines related to spirits and dreams. Nevertheless, most of the traditional healers wanted to supplement their local and traditional knowledge on plant medicines with modern scientific knowledge. In other words, they wished to learn what modern science had to offer about plant remedies that they themselves had extensive, but a different kind of knowledge about. However, most of the healers were disappointed, because they had experienced bioprospecting as transfer of their knowledge to the American company, but with limited transfer of knowledge the other way.

Third, the healers hoped that bioprospecting could contribute to Tanzanian society's increased acceptance of traditional healing and to raise the social status of healers. In Tanzania, the degree of tolerance of traditional healing and recognition of traditional healers as positive contributors to health services has varied considerably over time. Bioprospecting implies that scientists from the modern healthcare sector in the large city of Dar es Salaam as well as from the USA visit rural villages to seek out traditional healers and learn about their plant medicines. This is noticed in the local communities.

Fourth, after their experiences with Shaman, traditional healers in the area had devised a strategy to be used should there be another chance to take part in bioprospecting. Instead of naming or showing samples of the plants involved, the healers would give the bioprospectors unrecognizable plant extracts. The healers felt that should any of their plant extracts have promising properties, they would then stand stronger in negotiations with bioprospectors.

Shaman presented the collection expedition as an outstanding model for cooperation and local participation. However, we found that the cooperation took place on an unequal basis. Shaman alone determined the content of the cooperation and payment. The traditional healers did not have knowledge about other companies they could choose to cooperate with or the terms that they might have a chance to obtain in negotiations.

Shaman's narrative included a description of meetings between the company and local partners in which the company asked for and obtained informed consent to conduct bioprospecting. Moreover, at each collection location, the bioprospectors arranged a meeting to invite suggestions of worthy causes to which the company could make donations. After the expedition, the company itself decided the donations to be made. For several reasons this is problematic to regard as informed consent. First, the eagerness of the traditional healers to learn about the knowledge of modern science about their herbal remedies seemed not to have been an issue that the company had grasped. Second, the local participants did not have information about the value of the knowledge they gave the pharmaceutical company. Third, neither the size of immediate donations nor the extent of new donations in case of elaborations of commercial product was made known in advance to the local participants. Furthermore, this cannot be considered as actual negotiations, because donations were not discussed before the end of the bioprospecting activity (Svarstad 2003).

In the following years, the Tanzanian state did not put forth laws or regulations to ensure national and local income from bioprospecting. This means, for instance, that knowledgeable traditional healers remained more or less without rights. The most they could hope for was to be visited by bioprospectors such as Shaman, who would voluntarily give them some ‘gifts’.

In the wake of the Shaman bankruptcy, its patents were sold to another company. None of the patents sold seemed to be related to this expedition. In the biopiracy discourse, patents associated with bioprospecting are regarded as unequivocal negative for a source country that has biological diversity and related traditional knowledge. However, we found no basis for that conclusion. Patents may in fact ensure that a source country of genetic resources, and local contributors of traditional knowledge, can derive income from new medicines. Such patents prevent other companies from exploiting the traditional and modern knowledge without paying for its use. A prerequisite for patents benefiting source countries and local actors in this way is, however, that source countries have national regulations to enable such agreements (Svarstad et al. 2000).

Traditional healers might be able to negotiate to their advantage with individual bioprospectors through their secrecy on plant types. But to realize the full potentials for benefits, their home countries need to have relevant laws and procedures in place. However, many countries have not established such institutions. An activity such as bioprospecting might results in civil servants exploiting opportunities for extra income for themselves or their agencies rather than acquiring a source of revenue for the nation state. According to Bayart (1993), many African bureaucracies work in line with such ‘politics of the belly’. In 1998, we visited government agencies in Tanzania to ask about the correct procedure for bioprospecting. At several of these agencies, we were told that the correct procedure was to pay a fee to the particular agency we visited.

Shaman’s narrative on the collection expedition in the Kilimanjaro region highlights the participation of local women and thereby that their approach implied gender equity. This came about by inviting not only traditional healers, who mostly are men, but also traditional midwives, all of whom are women. However, in our interviews with traditional midwives, we found that almost none of them had taken part in the meetings between the bioprospectors and local participants where the local benefits were discussed. Traditional midwives we interviewed who had worked with Shaman in the collection of samples of medicinal plants were not informed about the company’s contribution to building a centre for the organizations of which they were members. In other words, our research questioned the Shaman narrative contention that women participated equally with men. At the same time, our investigation also revealed that the claim that the women who took on tasks under the label of ‘traditional birth attendants’ did not possess much knowledge of plant medicines, and instead of a traditional role, their work was based on first aid training within the modern health sector related to births (Svarstad 2003).

In addition, the Shaman narrative included a listing of ten authors of the research paper dealing with the collection expedition (Richter et al. undated). Of the ten, two (Richter and Ali) worked for Shaman. Seven were Tanzanian participants in the expedition, and one was a prominent traditional healer in the visited area. We interviewed five of the Tanzanians listed as authors to find that none of them had contributed to the manuscript. One did not know about the manuscript at all. An American consultant living in Tanzania had organized the expedition. He had contributed extensively to the paper, but was not included in the list of authors. Shaman had sought to use the list of authors to convey the impression of equality in the cooperation between the company and 'authentic' collaborators in Tanzania. Finally, the manuscript was never published, and probably this was due to disagreements between Shaman and some of the 'authors'.

In conclusion, what we learned in the field about the case showed that there was no basis for confirming Shaman's win-win narrative. Was this then an example of biopiracy supporting the idea that bioprospecting should generally be rejected as an economic activity in the Global South? We do not believe that bioprospecting provides a good source of potential income for Tanzania or for traditional healers. As discussed above, the lack of income is not due to patenting, as contended in the biopiracy discourse. Nor in this case can it be said to be due to lack of ethics on the part of the bioprospector, who after all donated more than they needed to in order to gain access to biological resources and traditional knowledge. On the other hand, a principal cause why the win-win narrative did not materialize—in fact in accordance with the win-win discourse—is that the Tanzanian public administration had not established the necessary institutional framework to enable and regulate this type of economic activity.

Even though the pharmaceutical industry remains committed to bioprospecting, this industry currently seems to have become less interested in traditional knowledge. In looking at bioprospecting around the world for the last few years, it is almost impossible to find incidents in which this activity provided any substantial income for people who have contributed with traditional knowledge. Some researchers contend that there are two exceptions. One is that the Kani tribes of the state of Kerala in India are to receive a considerable share of the income from a company that sells a natural medicine based on their knowledge. The other is that the San people of southern Africa have negotiated with the multinational companies Phytoharm and Unilever for their share of the income from a weight reduction extract made from the Hoodia plant. The San are an indigenous people of an arid area who traditionally have used Hoodia as an appetite suppressant (Wynberg and Laird 2007; Wynberg et al. 2009).

Thus, income from bioprospecting has not provided the treasure trove for tropical countries that many delegates to the international biodiversity negotiations in the 1990s hoped for. An important reason is that corporations do not seem to depend on continuous supplies of biodiversity samples for their development of new products. Another reason is the international campaigns against the activity based on the biopiracy discourse. This may have avoided many cases of exploitation, but at the same time, it has also prevented opportunities for benefits for possessors of local and traditional knowledge.

3.8 From Hollywood to the Norwegian Mountains

We will now move on to present a case of bioprospecting that took place in Norway as early as 1969. This case is interesting for two reasons. First, because it provides insights to the political economy of bioprospecting. This is an activity that usually takes many years from biological samples are collected, until a new product is developed, becomes available on the market, and makes substantial revenues.

Second, it also provides an interesting case about how a narrative may be created and established by media as a truth, and this is what we will highlight here.

Earlier in this chapter we have seen how the *Medicine Man* film captured elements of both of the two types of discursive narratives on bioprospecting that have been prominent since around 1990. Contrary to the Hollywood presentation of bioprospecting, Norwegian media have given one-dimensional presentations of biopiracy. This has happened even in the case discussed below where our research was the original source of knowledge, which provided a very different narrative.

The biologist Hans Peter Frey worked at Sandoz, a Swiss company that later became part of Novartis, the largest pharmaceutical company of recent decades. In interviews with Frey, he told us that in 1969 he and his wife had taken an autumn holiday in Norway. In Oslo, they rented a car that they drove over the mountains to Bergen. When passing the Hardanger mountain plateau they stopped along the road several times to take photos of the scenery. Frey told us that at each stop he took the opportunity to collect a soil sample. Back in Switzerland, the samples went through a comprehensive process Sandoz had devised in its search for medicinal properties of biological material. One of the samples from the Hardanger Plateau contained a microfungus named *Tolypocladium inflatum*, and this fungus was found to produce cyclosporin, a biochemical that prevents immune reactions after organ transplantations.

From 1978 to 1998, Sandoz and its successor Novartis took out many patents associated with cyclosporin, including 15 American patents. In 1983, the first medicine based on cyclosporin was introduced on the market. By 1997, it had become Novartis' most profitable pharmaceutical product, with a market of more than one billion dollars that year (Svarstad et al. 2000).

For years, Norwegian biologists and conservationists have pointed to the 'Hardanger Plateau fungus' as an example of the importance of protecting biological diversity (for example Høyland and Ryvarden 1990; WWF Norway 1998). In this narrative, the argument is that medicines crucial for humanity can be found in biodiversity everywhere, even in barren Norwegian mountains. Thus, protected areas are important, although the sample behind the medicine was actually collected outside the Hardanger Plateau National Park.

In the late 1990s, the microfungus became the object of a biopiracy narrative in which biological patents were held to be the principal means of exploitation. The magazine *Folkevett* ('Common Sense'), published by The Future in Our Hands, Norway's largest environmental organization, wrote that:

> The patent wave is hitting us hard. Norway loses millions every year because a foreign company has patented a medicine derived from a fungus from Norwegian nature. Developing countries lose even more, when pharmaceutical companies grab a piece of their nature and patent so-called biological innovations without giving anything back to the source country (*Folkevett* no. 4 1998. Our translation from Norwegian).

With *Folkevett* as the source, the narrative was repeated on the leading Norwegian TV news (27.02.1999). Likewise, *Aftenposten*, the country's largest newspaper leaning politically towards centre-right wrote (16.06.2000):

> The foreign company Hoffman LaRoches took out a patent, when researchers found a microorganism on the Hardanger Plateau, that they developed into a medicine that prevents the human body from rejecting transplanted organs, such as a heart. Because of the patent, Norway does not recoup any of the billions that the medicine brings in each year. (Our translation from Norwegian)

The references here to a wrong company and just one patent are symptomatic for the lack of weight on empirical knowledge when accounts of biological patents and biopiracy have been presented in Norwegian media. Through our research, we have found that patents in this case did not constitute a cause of exploitation. Norway may be seen as having lost income in this case, but this is due to other reasons. Like the rest of the world, Norway did not have laws that regulated access to genetic resources, when the sample of the microfungus was taken to Switzerland in 1969. With such regulations, we have estimated that Norway would have had a realistic claim against Novartis for about 24 million dollars per year (Svarstad et al. 2000).

Novartis' patents provide payment for the expense of developing the success medicine. During the years that the patents are in force, other companies are barred from using the findings without paying for the research. If Norway had appropriate legislation at the time and a contract with Novartis on a share of the income from such a finding, the patents could have ensured benefits to the Norwegian state (Svarstad et al. 2000; Svarstad 2002). This may not have been essential for Norway, which in the 1970s developed its fossil fuel industry, but for low income countries such income, in coherence with the Biodiversity Convention, could imply useful additions to national budgets. If so, the focus would have had to be moved from patenting to laws and procedures for bioprospecting.

So, what was the source of the reports in Norwegian media that patenting prevented Norway from benefitting from the Hardanger Plateau fungus? We were surprised to find that this was actually our own research, although our research conclusions were not presented at all, but instead they were replaced with unsubstantiated speculations. The media reports presented our calculations of loss of income, but ignored our assessment of the effect of patenting and instead reproduced the biopiracy narrative. Although we made the first narrative producer (*Folkevett*) aware of this before their article was printed, they chose to print it without our corrections, and their presentation constituted the source for the other media.

One may ask why patenting, for instance in this example, is given a central role as cause of exploitation in the biopiracy discourse, without empirical grounds and in contrast to research. In order to understand this, it is important to notice the way the biopiracy discourse is produced according to a similar pattern as a group of what we may label *radical discourses* on various topics on environment and development. Several of these discourses focus on patenting as the cause of unfairness and exploitation. Sometimes this view may be correct, but in the case of the Hardanger Plateau fungus, we observed how the claim was made without evidence and even despite contrary evidence. Thus, sometimes claims are transferred from one discourse to another without proper examination of whether or not the claim is supported by evidence (Svarstad 2002).

Leading discourses are those discourses that are produced by powerful actors, implying that they have consequences for policies, laws and practices. We find it important to examine such discourses and compare them to empirical studies of the topics that the claims of these discourses focus on. In several of the following chapters, we will make such comparisons between discourses and critical empirical examinations. At the same time, we argue that it is important to study specific discourses in the context of other discourses in terms of the discourse order of a specific issue as well as clusters of discourse types.

3.9 Four Types of Discourses on Environment and Development

As we have seen, Shaman presented its bioprospecting in Tanzania in the form of a win-win narrative, whilst the case of the Norwegian fungus has been subject to the production of a biopiracy narrative. However, our own research in each of these cases found no basis for agreeing with either the win-win or the biopiracy narrative.

Discourses and their associated narratives help simplify reality, and this makes it easier to interpret issues and cases that otherwise may appear as complex and difficult to understand. The simplification facilitates action, but discourses also bar other possible actions (Dryzek 1997). Political ecologists who provide critical examinations of 'established truths', may bring forth knowledge and perspectives that can contribute to debates, policies and practices in directions not thought of previously. Contributing in this way presupposes familiarity with the discourses produced in a field as well as sound empirical knowledge of the questions at hand.

It also requires the will and possibility to ignore expectations, which we have observed and experienced ourselves, that researchers ought to deliver knowledge confirming the content of particular discourses. Such pressure seems to be increasing in many countries, as research is organized in ways that require the financial support of research programmes in which public agencies may delineate narrow discursive frames for the questions that researchers are allowed to address.

Table 3.1 Four discourse types on topics of environment and development

Discourse types:	Is conservation of natural resources considered vital?	Are needs and interests of local people considered vital?	Does the discourse type contain positive views on partnerships between local users and external actors?
Preservation discourse	Yes	No	No
Win-win	Yes	Yes (as a means)	Yes
Radical	Yes	Yes	No
Prometheus	No	Yes	Irrelevant

Source: The authors

Through comparing leading discourses on various topics in environment and development, we have identified four *discourse types* (Svarstad et al. 2008). We distinguish between these discourses, which we label win-win, radical, preservation, and Prometheus (see Table 3.1). The discourse types are based on studies conducted of discourse production on a limited number of topics. Thus, they should not be regarded as generalizations of the discourse orders for all themes on environment and development. Instead, they can be used as ideal types (Weber 1949), that form comparative bases for analyses of discourse production on topics in the same area.

In a study, we compared the findings of the analysis already done on bioprospecting with the production of leading discourses on climate change, deforestation and desertification (Adger et al. 2001).

On all the topics, we found variations of a *win-win discourse type*. The principal focus in this type of discourse is the conservation of natural assets, and this is in various ways linked to other goals, such as economic gains and losses for groups and/or countries. Local participation in decisions and economic gains are often important means in conservation initiatives. The win-win discourses are based on a positive attitude for cooperation between local and external actors. The external actors may come from the same country or from abroad. They may be private companies as well as conservation organizations or public environmental agencies. In the next chapter, we will see how a win-win discourse plays a key role in the governance of protected areas.

A win-win discourse can to various degrees be linked to the political and economic discourse of neoliberalism. However, this does not necessarily mean that win-win discourses are based on the idea of unregulated markets, but rather that the discourse points to the importance of partnerships linked to the establishment and regulation of new markets.

On all the four investigated topics, Adger et al. (2001) identified what may also be called a *radical discourse type*. The biopiracy discourse belongs here. Radical discourses focus primarily on consequences of external interventions for local resource users. In contrast to win-win discourses, radical discourses do not regard partnership between local and external actors generally as positive. Instead, there is a focus on unequal power relations, so that local actors have little influence. Moreover, radical discourses are concerned about imbalances in the distribution of economic costs and benefits from such partnerships between various external actors on one hand, and local actors on the other. In addition, attention is often put on consequences for the most disadvantaged as well as for indigenous people and other marginalized groups.

Radical discourses regard sustainable governance of land and natural resources as important. Local actors are considered capable of managing land and resources sustainably, provided that they are not subjected to external pressure or interference from markets, businesses or conservationists. In the next chapters we will see how the precursor of a radical discourse on protected areas is in accordance with the radical discourse type.

Among the roots of the academic contributions to radical discourses, we find ideas of self-reliance in developing countries and dependency theory that were strong in the 1970s and 1980s. Since the 1990s, some radical discourses have acquired elements both from the globalization resistance movement as well as from post-structuralism. In the following chapters, we provide examples showing that radical discourses continue to be produced on various topics.

The third type consists of *preservation discourses* where the focus is on maintaining natural assets such as areas and species. This discourse does not include concern for local resource users. Consequently, it is not seen as important to establish partnerships between external and internal actors. It is the national and international organs that must ensure conservation, and local actors are regarded as impeding such conservation.

This discourse type reflects viewpoints held by many of the earliest conservationists. Preservation discourses comprise the precursors of many win-win discourses on environment and development. Some have argued for a revival of preservation, with its defence of conservation combined with neglect of local livelihoods (Hutton et al. 2005). At the same time, in many cases preservation can be argued to constitute the basis for *practises* of conservation in poor countries, even though practitioners may tend to tell about their activities in line with the win-win discourse (see ▶ Chap. 4).

The fourth and last of the discourse types we have found on environment and development is what we, in line with Dryzek (1997), call a *Prometheus discourse type*. In Greek mythology Prometheus stole fire from Zeus and thereby enabled people to exploit nature. In this discourse type it is seen as unnecessary to worry about conservation. Every problem that arises—degradation of biodiversity, climate change or any other problem—will be solved by new technologies. From the Industrial Revolution up to the last few decades, modern societies have been dominated by a Promethean understanding (Dryzek 1997).

Today Promethean claims and discourses play a minimal role related to many environment and development topics. There are, however, important exceptions.

There are, for instance, still climate sceptics who reject the dominating views by scientists that serious anthropogenic climate change is taking place.

The identification of the four discourse types on topics about environment and development does not imply that we argue that leading discourses on other topics have to be in consistence with these types. For each topic, an empirical examination has to be conducted in order to identify the discourse order and those discourses that influence politics and practices. Moreover, discourses are not static. Their continuations as well as changes rely upon their co-production of many actors, and crucial aspects of the contexts of production may change over time. There may also be important differences between discourses produced on a topic at the global level and national varieties of these. It is also important to keep in mind that even though we distinguish here between four clearly different types of discourses, it may sometimes be difficult to set clear boundaries between specific discourses. There may be intermediate positions containing elements of several discourse types.

Nevertheless, we think that the four discourse types often may constitute a useful basis for comparison in order to identify discourse orders, and first of all the leading discourses, of particular environment and development topics. When analysing a material of statements about a topic, this would involve comparisons of similarities and differences with the discourse types.

Methods: How Can We Study Discourses?

First, it usually makes sense to start by identifying leading discourses on a theme. This can be done in a range of different ways. When beginning the research on a selected topic, one may start with a literature search to identify relevant studies that have already been carried out by other scholars about leading discourses on the topic. Such studies may be suitable to use as a point of departure for one's own empirical study based on critical realism (see ▸ Chap. 1), where claims in one or more of the leading discourses are discussed in comparison to one's own findings. Key claims may be about practices, results and impacts.

One may also choose to carry out an independent discourse analysis, because of a lack of relevant discourse analyses, or because existing studies are dated making it necessary to examine whether the discourse universe on the topic has changed.

We argue that discourse analyses that merely describe discourses usually have limited value in critical studies if it is not combined with one or more elements of realism where discursive claims are examined, or the use of discursive power is analysed.

Primary sources in discourse analysis may consist of media articles, presentations on websites, reports and documents from businesses, organizations or government departments. Radio and TV programmes as well as images may also be useful sources, as well as one's own interviews with key actors, and for this chapter's study of bioprospecting discourses, observations of conferences and international negotiations provided valuable insight and also journal articles written by bioprospectors. Sometimes there may be a few crucial sources that give clear presentations of leading discourses on a topic, while other times it may be necessary to examine a large number of sources.

We recommend basing the analysis of collected data on a coding sheet, where the sheet first is prepared before the analysis of the sources, and thereafter further elaborated along with the coding. As in other analyses of qualitative data, it is possible to use either a paper sheet or one of the available software programmes designed for qualitative data analysis.

We suggest starting the elaboration of the coding sheet by drawing on academic publications from other discourse analyses to make a list of features of leading discourses that one might be expected to find. This provides a deductive approach. For discourses on environment and development, the content in ◘ Table 3.1 may be useful as a start. In addition, theories are also often useful sources for questions about what discourses may contain in terms of claims about a topic. When going through and coding the data, one may register relevant parts of the texts and make notes about whether or not the various aspects are in line with, similar to, or in contrast to elements on the coding sheet.

During the coding, one should also look for unexpected aspects that may constitute important features of a particular discourse. This constitutes an inductive approach. Such findings should be added to the coding sheet and thereby included in the elements that one looks for also in other sources. Thus, in new rounds of coding, the relevance of the added elements on the coding sheet is included in the examination.

Sometimes it is quite straight forward to identify different discourses, in particular when they are polarized. Often, however, discourses are tangled together in complex ways, where some elements are shared between different discourses. Where to draw boundaries between discourses may in such cases not be obvious. As in other studies, the discourse analyst should discuss which findings form part of clear patterns, and where there are deviations from these patterns.

As discursive power is an important form of power, a discourse analysis in political ecology would also include an analysis of the actors involved in producing and reproducing the discourse as well as the practical implications of the discourse in terms of impacts on society.

An example of such an impact is presented in ► Chap. 6, where we show how 'international cost-effectiveness' is a key aspect in the leading discourse on climate change mitigation in Norway. This has led to outsourcing of climate mitigation measures to other European countries as well as to low-cost countries in the Global South and legitimized lack of climate action in Norway. This discursive element can be traced back to the lobbying by the fossil fuel industry around 1990, and can be seen as a successful exercise of discursive power.

Conclusions

In this chapter we have presented the topic of bioprospecting as an example of how some political ecology perspectives can be applied. Elements of all of the ten syntheses presented in ► Chap. 1 have been applied to some extent, and with a particular emphasis on synthesis five between realism and social constructivism. We provided examples of realism first by a short presentation of the political economy of the economic activity of bioprospecting, and then through our own studies of how bioprospecting has been carried out in two cases. Moreover, we presented a framework of social constructivism through the investigation of discourses and narratives on topics related to environment and development. The approach has been derived from the study of bioprospecting, and we consider bioprospecting as a useful example for its introduction also because the extensive polarization between the two leading discourses on bioprospecting makes it a relatively simple example. In the following chapters, we will provide more examples of combined studies of realism and social constructivism, and we also go more into other elements of the ten syntheses.

? Questions

1. Find a case of bioprospecting that either has taken place in your country or has been conducted by companies or scientists from your country (or another country you are interested in). Search for texts about the case from any type of source. Compare the presentation of bioprospecting in the texts with the win-win and biopiracy narratives.
2. Discuss what aspects you would examine if you were to make a study of this case based on realism.
3. Discuss the discourse concept presented in this chapter in comparison to other concepts such as ideology, culture, paradigm and theory.
4. Try to identify environmental issues in your country where a hegemonic discourse dominates.

References

Adger, W.N., T.A. Benjaminsen, K. Brown, and H. Svarstad. 2001. Advancing a political ecology of global environmental discourses. *Development and Change* 32 (4): 681–715.
Aristotle. 1996. *Poetics*. Harmondsworth: Penguin.
Bayart, J.-F. 1993. *The state in Africa: Politics of the belly*. New York: Longman.
Bell, J. 1997. Biopiracy's latest disguises. *Seedling* 14 (2): 2–10.
Brush, S. 1999. Bioprospecting the public domain. *Cultural Anthropology* 14: 535–555.
Carlson, T.J., M.M. Iwu, S.R. King, C. Obialor, and A. Ozioko. 1997. Medicinal plant research in Nigeria: An approach for compliance with the convention on biological diversity. *Diversity* 13: 1.
Castree, N. 2003. Bioprospecting: From theory to practice (and back again). *Transactions of the Institute of British Geographers* 28 (1): 35–55.

CBD. 2007. *Analysis of gaps in existing national, regional, and international legal and other instruments relating to access and benefit sharing.* Secretariat of the Convention on Biological Diversity, draft for peer review prepared for the Ad-Hoc Open-Ended Working Group on Access and Benefit Sharing, Fifth Meeting, Montreal, 8–12 October.

Chinnock, J.A., M.J. Balick, and S.C. Sanchez. 1997. Traditional healers and modern science – Bridging the gap: Belize, a case study. In *Building bridges with traditional knowledge*, ed. A. Paul, D. Wigston, and C. Peters. New York: New York Botanical Garden Press.

Coalition Against Biopiracy. 2008. Captain hook awards for biopiracy. Lesedato: 11.05.2010. http://www.captainhookawards.org/.

Dryzek, J.S. 1997. *The politics of the earth. Environmental discourses*. Oxford: Oxford University Press.

Eisner, T. 1989. Prospecting for nature's chemical riches. *Issues in Science and Technology* 6 (2): 31–34.

———. 1991. Chemical prospecting: A proposal for action. In *Ecology, economics, and ethics: The broken circle*, ed. F.H. Bormann and S.R. Kellert. New Haven: Yale University Press.

Elliott, J. 2005. *Using narrative in social research. Qualitative and quantitative approaches*. Los Angeles, London, New Delhi, Singapore: SAGE Publications.

Fairclouch, N. 1995. *Media discourse*. London: Edward Arnold.

Foucault, M. 1970. *The order of things: An archaeology of the human sciences*. New York: Random House.

———. 1979. *Discipline and punish: The birth of the prison*. Harmondsworth: Penguin Books.

———. 1988. *Madness and civilization: A history of insanity in the age of reason*. New York: Vintage.

Hajer, M. 1995. *The politics of environmental discourse*. Oxford: Oxford University Press.

Høyland, K., and L. Ryvarden. 1990. *Er det liv er det sopp. Sopp i miljø og kulturhistorie*. Oslo: Fungiflora.

Hutton, H., W.M. Adams, and J.C. Murombedzi. 2005. Back to the barriers? Changing narratives in biodiversity conservation. *Forum for Development Studies* 32 (2): 341–370.

Johansson, A. 2005. *Narrativ teori och metod. Med livsberättelsen i fokus*. Sverige: Studentlitteratur.

Juma, C. 1989. *The gene hunters. Biotechnology and the scramble for seeds*. London: Zed Books Ltd.

King, S.R. 1994. Establishing reciprocity: Biodiversity, conservation and new models for cooperation between forest-dwelling peoples and the pharmaceutical industry. In *Intellectual property rights for indigenous peoples. A sourcebook*, ed. T. Greaves. Oklahoma City: Society for Applied Anthropology.

Kloppenburg, J., and S. Rodriguez. 1992. Conservationists or corsairs? *Seedling* 9 (2–3): 12–17.

Lyotard, J.-F. 1997. *The postmodern condition: A report on knowledge*. Minneapolis: University of Minnesota Press.

Martinez-Alier, J. 1996. The merchandizing of biodiversity. *Capitalism, Nature Socialism* 7: 37–54.

Mateo, N. 2000. Bioprospecting and conservation in Costa Rica. In *Responding to bioprospecting: From biodiversity in the south to medicines in the north*, ed. H. Svarstad and S.S. Dhillion. Oslo: Spartacus.

Mooney, P. 2000. Why we call it biopiracy. In *Responding to bioprospecting: From biodiversity in the south to medicines in the north*, ed. H. Svarstad and S.S. Dhillion. Oslo: Spartacus.

Moran, K. 1997. Returning benefits from ethnobotanical drug discovery to native communities. In *Biodiversity and human health*, ed. F. Grifo and J. Rosenthal. Washington, DC: Island Press.

Neimark, B. 2017. Bioprospectig and biopiracy. In *International encyclopedia of geography: People, the earth, environment and technology*, STED, ed. D. Richardson, N. Castree, M.M. Goodchild, A. Kobayashi, W. Liu, and R.A. Marston. Chichester: Wiley-Blackwell.

Parry, B. 2000. The fate of the collections: Social justice and the annexation of plant genetic resources. In *People, plants, & justice. The politics of nature conservation*, ed. C. Zerner. New York: Columbia University Press.

Portillo, Z. 1999. *Biopiracy: A new threat for the Amazon*. Inter Press Service, Roma. Distributed January 6 by GRAIN's listserver BIO-IPR.

RAFI & Cultural Survival Canada. 1997. Sangre de drago. *BioPirates*, Log # 4. Rural Advancement Foundation International (RAFI) and Cultural Survival Canada.

RAFI (Rural Advancement Foundation International). 1994. *Conserving indigenous knowledge: Integrating two systems of innovation. An independent study*. New York: United Nations Development Programme.
RAFI. 1995. Biopiracy Update: A Global Pandemic. *RAFI Communique*, September/October 1995.
Reid, W.V., S.A. Laird, C.A. Meyer, R. Gámez, A. Sittenfeld, D.H. Janzen, M.A. Gollin, and C. Juma. 1993. *Biodiversity prospecting: Using genetic resources for sustainable development*. Baltimore: World Resources Institute.
Reyes, V. 1996. The value of sangre de drago. *Seedling* 13 (1): 16–21.
Richter, R.K., et al. Undated. *Mutualism: An ethnomedical research collaboration between Shaman Pharmaceuticals and the Institute of Traditional Medicine in Tanzania*. Unpublished draft article. San Fransisco: Shaman Pharmaceuticals.
Roe, E. 1991. Development narratives, or making the best of blueprint development. *World Development* 19 (4): 287–300.
———. 1995. Except-Africa: Postscript to a special section on development narratives. *World Development* 23 (6): 1065–1069.
———. 1999. *Except-Africa. Remaking development, rethinking power*. New Brunswick: Transaction Publishers.
Shaman Pharmaceuticals. 1997a. *Shaman Pharmaceuticals. 1996 annual report*. San Francisco.
———. 1997b. *Products in development & research focus*. San Francisco: Shaman Pharmaceuticals.
Shiva, V. 1997. *Biopiracy. The plunder of nature and knowledge*. Boston: South End Press.
———. 2007. Comparative perspectives symposium: Bioprospecting/biopiracy: Bioprospecting as sophisticated biopiracy. *Journal of Women in Culture and Society* 32 (2): 307–313.
Svarstad, H. 2000. Reciprocity, biopiracy, heroes, villains and victims. In *Responding to bioprospecting. From biodiversity in the south to medicines in the north*, ed. H. Svarstad and S.S. Dhillion. Oslo: Spartacus.
———. 2002. Analysing conservation-environment discourses: The story of a biopiracy narrative. *Forum for Development Studies* 29: 1.
———. 2003. *Bioprospecting: Global discourses and local perceptions – Shaman Pharmaceuticals in Tanzania*. SUM dissertation and thesis 6/03, Center for Development and the Environment, University in Oslo.
———. 2005. A global political ecology of bioprospecting. In *Political ecology across spaces, scales and social groups*, ed. S. Paulson and L. Gezon. New Brunswick: Rutgers University Press.
Svarstad, H., H.C. Bugge, and S.S. Dhillion. 2000. From Norway to Novartis: Cyclosporin from Tolypocladium inflatum in an open access bioprospecting regime. *Biodiversity and Conservation* 9: 1521–1541.
Svarstad, H., L.K. Petersen, D. Rothman, H. Siepel, and F. Wätzold. 2008. Discursive biases of the environmental research framework DPSIR. *Land Use Policy* 25 (1): 116–125.
SynBioWatch. 2016. The 2016 Captain Hook Awards for Biopiracy – submit your nominations now! http://www.synbiowatch.org/2016/11/the-2016-captain-hook-awards-for-biopiracy-submit-your-nominations-now/. Visited 14 May 2020.
United Nations. 1992. *Convention on biological diversity*. Rio de Janeiro. http://www.cbd.int/convention/convention.shtml.
Weber, M. 1949. 'Objectivity' in social science and social policy. In *The methodology of the social sciences*, ed. E.A. Shils and H.A. Finch. Glencoe: The Free Press of Glencoe.
WWF Norway. 1998. Soppen fra Hardangervidda. *Verdens Natur 2*.
Wynberg, R., and S. Laird. 2007. Bioprospecting – Tracking the policy debate. *Environment* 49 (10): 20–32.
———. 2009. Bioprospecting, access and benefit sharing: Revisiting the 'grand gargain'. In *Indigenous peoples, consent and benefit sharing: Lessons from the San-Hoodia case*, ed. R. Wynberg, D. Schroeder, and R. Chennells. Dordrecht: Springer.
Wynberg, R., D. Schroeder, and R. Chennells, eds. 2009. *Indigenous peoples, consent and benefit sharing: Lessons from the San-Hoodia case*. Dordrecht: Springer.

Conservation Discourses Versus Practices

Contents

T. A. Benjaminsen, H. Svarstad, *Political Ecology*,
https://doi.org/10.1007/978-3-030-56036-2_4

Trailer

September 2009, Longido District, Tanzania: We are driving along a bumpy, dusty road, which hardly consists of more than two tire tracks through a dry savanna landscape on the border to Kenya. This is the land of Sinya, a Maasai village, close to a recently established community-based conservation area called Enduimet Wildlife Management Area. Occasionally we see wildlife such as zebras, giraffes, antelopes and elephants. Because of a severe drought this year, we see a lot of dead livestock. The Maasai pastoralists in this region have lost many cattle in the last few months. There is also some dead wildlife around.

We see the dying elephant from a long distance in the open, barren landscape. It is still breathing, but dies a few hours later. However, it is not the drought that killed it. The elephant had wandered around for a couple of days with a long Maasai spear in its head. When we see it, the spear had just been removed by veterinarians. There are also uniformed rangers with rifles at the site, along with representatives from the District authorities' Wildlife Office.

The elephant was not killed by poachers hunting for ivory. The drought is a significant factor behind what has happened. Lack of water has led to competition between humans and elephants in the area. Rumour had it this elephant had arrived at a waterhole and had not managed to scare away all the cattle drinking there. The thirsty elephant then trampled a cow to death. This led some Maasai herdsmen to attack the elephant and place a spear in its head. This is an example of the struggle for water between elephants and humans, which tends to take place in this area during droughts. Many small-scale farmers in the area are also frustrated because the elephants often come and raid their crops. Still, however, attributing what happened to the drought alone would be too simplistic.

Overview

In contrast to drought as an explanation, one may argue that the reason the elephant died is because the number of elephants in the area had increased before this incident (Moss 2001; Kioko et al. 2006), and that this is related to the increase in different types of protected areas and reserves. In this area, and in Tanzania in general, the extension of conservation areas in different forms has continued to increase. According to official figures, 36% of Tanzania was covered by various conservation areas in 2007. Since then, new community-based forest and wildlife conservation areas have been established. In 2015, Wildlife Management Areas alone were estimated to cover 13% of the land area (Bluwstein and Friis Lund 2018). This means that totally around half of Tanzania's land mass is conserved in one way or another, which again leads to a reduced area in which people may keep livestock and farm the land.

People have also experienced an increase in crop damage caused by elephants. In addition, conservation in Tanzania is most of the time carried out with modest participation from local communities who often feel marginalized politically as well as economically through loss of access to land and natural resources, whether the conservation is called 'community-based' or not (Neumann 1998; Brockington 2002; Goldman 2003; Igoe and Croucher 2007; Benjaminsen and Bryceson 2012; Benjaminsen et al. 2013; Bluwstein 2018).

This feeling of being disempowered may lead to increased resistance against conservation and result in wildlife such as elephants being attacked. The number of elephants has, in recent years, gone down in Tanzania. Organized poaching and smuggling of ivory to illegal markets in South-East Asia are usually blamed for reduced elephant populations. It is, however, often forgotten that the top-down manner in which most conservation is carried out leads to resistance against conservation efforts from governments and international organizations. This again may make it easier to recruit local people into poaching networks (Hübschle 2017). In addition, elephant killings may also be seen as a message of protest against conservation practice (Mariki et al. 2015).

In fact, resistance against conservation may follow different tactics including nonviolent, militant or violent, discursive and formal-legal (Cavanagh and Benjaminsen 2015). The first may be associated with various forms of 'the weapons of the weak' (see ► Chaps. 1 and 2). The second can be in the form of armed rebellion or other forms of open violent resistance such as attacking rangers. Discursive resistance will for instance be by attempting to influence policies and the public debate in general. This usually happens through local leaders or local people trying to influence outsiders (politicians, journalists, NGO representatives, researchers) to speak for communities. Formal-legal resistance involves navigating the law enforcement and justice system, which means administrative complaints and taking cases to the courts.

The example with elephant killings definitely represents a form of violent resistance, but at the same time, it may also be seen as an application of the 'weapons of the weak'. People's efforts to present the death of the elephants in interviews with us as an accident, is in line with this theory, as well as the villagers' collective decision of not exposing the names of people responsible for elephant killings (Mariki et al. 2015). Furthermore, Scott (1989, p. 56) argues that 'when the act of everyday resistance is meant to be noticed—meant to send a signal—as in the case of arson or sabotage, then the resisters take special care to conceal themselves often behind a facade of public conformity'.

Hence, rural people in Africa killing wildlife in frustration and as protests against conservation is contrary to the picture often given by conservation organizations, tourist agencies and African governments of conservation as not only benefitting wildlife, but also local people. So, how has this harmonious image of African conservation been established?

To try and answer this question, we believe it may help to make use of a discourse perspective with an analytical distinction between *discourses* and *practices*. This perspective enables the disclosing of differences between the way in which powerful actors present their own practices and observations of these practices. In this chapter we focus on such discrepancies between what various actors *say* they do and what they actually *do*.

We proceed by a brief discussion of how we see the differences between discourses and practices. Then we introduce some main discourses on protected areas before we discuss conservation practices through the presentation of one case study from South Africa and one from India.

4.1 Discourses and Practices

In ► Chap. 3, we defined a discourse as a socially shared perspective on a phenomenon. Each discourse is based on assumptions, claims and arguments. Discourses can be seen as lenses through which the subject is viewed, and discourses are created and maintained by social actors. Within each discourse, contributions have some similarity in content, and they may also be marked by use of the same types of expressions such as certain metaphors and manners of relaying specific examples.

Dryzek (1997) argues that a discourse may have useful features for society by providing a key to interpret environmental issues and thereby facilitating action. At the same time, discourses also make their proponents blind from seeing alternative interpretations and actions. Thus, some discourses create important frameworks for interpretation of specific issues, and political decisions and ways of handling these issues can be seen as results of these discourses. There are global actors who contribute particularly actively to the design of leading discourses. These actors thus practise discursive power.

We show in this chapter that there are a few discourses on the issue of conservation, which have, in various periods from the beginning of conservation in the late nineteenth century until today, been prevalent and dominated at a global level.

Whereas discourses consist of something that is *said* and *written,* we see *practices* as the actions that are *carried out*. Thus, conservation discourses are produced, while practices of conservation take place, for instance, in the specific ways protected areas are established and managed. Some may claim that these two dimensions are inseparable and should be treated as such. Absolute borderlines can obviously not be drawn between discourses and practices. Nevertheless, we argue that it is useful to make an analytical distinction between the two in order to show how and why the portraying of a specific practice by involved actors may differ substantially from the observations we have of the same practice. Thus, we show how some actors contribute to the production of one discourse, while at the same time taking part in a practice in compliance with a different discourse.

This approach implies that we are open to the possibility that some of the contributions to the production of a discourse may come from actors that sometimes deliberately tell about specific practices in ways that deviate from their own knowledge about these practices. These are actors who see an interest in portraying the practices in coherence with a particular discourse.

The approach we use here can be combined with the approach that Igoe et al. (2010) propose for studies on conservation based on Gramsci's terms 'hegemony' and 'historic bloc' (Gramsci 1971). An idea, or, as we will emphasise, a discourse on conservation, 'becomes hegemonic when it is so systematically and extensively promoted that it acquires the appearance of being *the only* feasible view of how best to pursue and implement conservation goals' (Igoe et al. 2010: 488). A historic bloc

> refers to a historic period in which groups who share particular interests come together to form a distinctly dominant class. The ideas and agendas of this class thus come to permeate an entire society's understanding of the world (Igoe et al. 2010: 489–490).

Since the first protected areas were established in the late nineteenth century, different discourses have been developed. The oldest is the discourse of fortress conservation. Today, policies on nature protection in the Global South are primarily dominated by the win-win discourse on protected areas, which again is contrasted by a critical discourse emerging from human and indigenous rights activists as well as political ecology.

4.2 Fortress Conservation as Discourse and Practice

In 1872, Yellowstone was established as the first national park in the USA and the world. The focus of the park was on the protection of nature and the only form of use permitted was tourism. Visitors were wealthy white city residents. John Muir, one of the founders of American conservation, said for instance that for purposes of conservation, all wilderness should be cleared of residents and reserved for the needs of urban people for recreation and spiritual refreshment (Dowie 2005).

The indigenous people who had resided in Yellowstone were therefore expelled, and their use of nature concealed. The park was presented as a wilderness area untouched by human influence, and indigenous names were replaced by nature-descriptive place names. Intruders were kept out by means of armed control. The US Army was given the responsibility of managing Yellowstone in 1886, and kept it until 1918 when a civil park management took over. However, the employees still used military-style uniforms, and soldiers were hired as wardens (Adams 2004).

This conservation practice at Yellowstone became a model for the subsequent establishment of protected areas throughout the world and formed the basis of the first discourse on what and how protected areas are and should be. This type of conservation is often termed Fortress Conservation because the protected areas are more or less defended like fortresses by uniformed, armed men. The permitted exceptions are tourism, natural science research and the production of nature documentaries.

This fortress conservation discourse later became important in the design of national parks in the colonized parts of the world, particularly in Africa and parts of Asia. Parc National Albert in The Belgian Congo (today this is the Democratic Republic of Congo and the park is called Virunga National Park) was established as the first national park in Africa in 1925, followed by Kruger National Park in South Africa in 1926. In the period between the Second World War and the liberation of the various African states around 1960, a large number of national parks were established in Africa.

Typically, 'wilderness' has often been perceived in opposition to 'civilisation' (Nash 1982). These were areas supposedly unaffected by modernization. Wilderness was far away from cities, roads and other signs of civilization. Today, 'wilderness' is one of the two strictest forms of protection within the IUCN's classification of protected areas used in international conservation.

Those involved in the production of the fortress conservation discourse include, first of all, state natural resource managers who have established and managed the protected areas as well as a group of big international non-governmental organizations (so-called conservation BINGOs). Many of the individuals involved are educated within biology, and more recently within specializations such as conservation biology and natural resources management. Natural scientists, who work at universities and research institutes, have also been key re-producers of this discourse.

4.3 The Win-Win Discourse on Protected Areas

From the late 1980s, the fortress conservation discourse was gradually overtaken by a competing discourse focusing on the role of local participation in conservation (Adams and Hulme 2001). This discourse consists of two main elements. First, it states that in order to achieve conservation it is necessary to let people in and around the protected areas participate in the management of these areas. Second, local communities should benefit from conservation, which will also lead to local support for the protected areas.

As in the fortress conservation discourse, the primary concern for the win-win discourse is to conserve biodiversity. However, it promotes an integration of interests of local people as a means to achieve conservation. Thus, the set-up involves aspects of benefit-sharing, compensation, and local participation, and the partnerships are argued to constitute win-win situations implying both environmental conservation and local development. Four main factors have made the win-win discourse on conservation hegemonic.

First, it draws upon the basic ideas that became influential through the Brundtland Report 'Our Common Future' of a win-win relationship between environment and development (World Commission on Environment and Development 1987). Second, from the 1980s, there was an increased pressure from development and human-rights-based activists and indigenous groups to change conservation practices in a more 'people-friendly' direction (Adams and McShane 1992; Bonner 1993). Third, the shift to the win-win discourse on protected areas also formed part of a more general shift within developmental policy, from a 'top-down' approach to a focus on 'participation' (Cooke and Kothari 2001; Adams 2009). Fourth, neoliberal economic policies have had great influence on conservation policy in general during the last couple of decades (Sullivan 2006; Heynen et al. 2007; Igoe and Brockington 2007; Brockington et al. 2008; Igoe and Sullivan 2009; Büscher 2010). The win-win discourse on protected areas is in line with neoliberal thinking by the emphasis on how markets for tourism are needed to finance conservation.

All the major conservation organizations, such as IUCN, WWF, the Nature Conservancy and Conservation International describe their activities as based on such a people-friendly strategy. Claude Martin, former Director General of WWF International, provided an example of a contribution to the win-win discourse from big conservation saying that

> » for a protected area to be successful, local communities and indigenous people must be adequately involved from the earliest planning stages. Local knowledge and traditional management practices are being included in management strategies, and a growing number of protected areas are being managed by local communities and indigenous people.

Photo 4.1 AWF billboard in northern Tanzania. (*Photo: Tor A. Benjaminsen*)

Furthermore, Martin argued that 'for the continued survival of the protected area, they (local communities) must share the benefits that arise from it' (Martin 2003).

In a similar spirit, the African Wildlife Foundation (AWF) called out its message 'Conserving Wildlife, Protecting Land, Empowering People' through a giant billboard placed close to Kilimanjaro Airport in northern Tanzania. The billboard depicted a person, presumably an African farmer or pastoralist, 'holding hand' with an elephant trunk (Photo 4.1).

Also, government officials within the state segment on natural resources in African countries contribute to the win-win discourse. An example of this is an address given by a representative of the Wildlife Division in Tanzania at the World Conservation Congress in Barcelona in 2008:

> In Tanzania we have a programme called "community-based conservation" where we involve the local people in conservation of protected areas and areas around their villages. … we have areas called Wildlife Management Areas where the local people plan their land uses in the areas, which include nature resource management and also other land uses. And from there they get benefits. So the important thing here is the benefit sharing with the local communities. The funds, which accrue from the hunting of animals, go to the local communities and part of it goes to the government… They also help to build schools and also hospitals for the rural communities, and therefore they also help to put their lifestyles into better shape (transcript from recording, World Conservation Congress, Barcelona, October 2008).

In addition to the active contributions to the win-win discourse by large conservation groups and government officials in Africa, various other actors have also adopted this discourse, such as multi- and bilateral development agencies, celebrities, owners of hotels and other tourism companies located next to or within protected areas, other corporations that donate money to conservation, as well as many scientists with various specializations who carry out research related to protected areas.

The win-win discourse on protected areas can be seen as an important element of what Igoe et al. (2010) following Sklair (2001) label as 'the sustainable development historic bloc' where networks of the above mentioned actors present markets, commodification and consumerism as key tools to ensure a combination of economic growth and biodiversity conservation.

4.4 Challenges to the Win-Win Discourse

Two very different challenges to the win-win discourse can be identified. First, there is what we will term a 'radical discourse on protected areas'. Here, the interests and rights of local communities are the primary concerns. Strong resistance is expressed against external top-down intervention in the form of conservation, and it is questioned whether the practices of local communities are a threat to sound environmental conservation. Furthermore, it is argued that practices labelled 'win-win' rarely involve real devolution of authority, but on the contrary lead to political and economic marginalization. Customary rights and historical use of land and natural resources are, in addition, often not recognized by states, which easily leads to local communities being overrun by the state or international organizations.

It is argued within this discourse that the financial benefits received by local communities are modest, and that these benefits must be compared to the costs borne by people who live where the area protection is established or extended. It is also claimed that protection of biodiversity tends to be prioritized over poverty alleviation if these two processes are seen to be in conflict with each other (e.g. Neumann 1998; Dzingirai 2003; Chapin 2004; Kepe 2004; Sullivan 2006; Igoe and Croucher 2007; Benjaminsen et al. 2008).

Finally, attempts at introducing community-based conservation may also work as a mechanism for dispossession of villagers' access to resource benefits from village land allowing conservation a foothold in these lands. This does not need to involve re-location and may happen through a gradual and piecemeal process or it may involve violence in other cases (Benjaminsen and Bryceson 2012).

The radical discourse is produced by actors such as local residents close to protected areas, their supporters including activists for indigenous groups and human rights activists, as well political ecologists and other critical scholars.

The other challenge to the win-win discourse consists of a type of renaissance of the old fortress conservation discourse (see Wilshusen et al. 2002; Brechin et al. 2003; Hutton et al. 2005). A number of preservation biologists and conservationists have claimed that when local communities are involved in the management of conservation areas, conservation is weakened in such a way that biodiversity is threatened (Kramer et al. 1997; Brandon et al. 1998; Oates 1999; Terborgh 1999; Sanderson and Redford 2003; Du Toit et al. 2004). These actors argue instead for a management, which is totally controlled by central agencies and based on scientific research. Concern for population growth in and around conservation areas is also used to reinforce the fortress conservation discourse (e.g. Veldhuis et al. 2019). The argument is that the numbers of people and livestock have become so high that community benefits represent a threat to conservation.

Despite these two challenges, the win-win discourse still constitutes the key perspective that is presented both by big conservation NGOs and by conservation authorities in most countries in the Global South. In the following sections, we present the *practices* identified in case studies from South Africa and India in order to compare these practices with what the win-win discourse says.

Namaqua National Park in South Africa

After the establishment of the new democratic South Africa in 1994, land reform was seen as an important means of fighting poverty and injustice. Namaqualand is a dry region in the north-western corner of South Africa with a population of around 70,000 and covering approximately 48,000 km^2. The population consists of ethnic Nama, people of mixed origin, and a minority of European origin. The latter group comprises approximately 20% of the population in this area. The former two categories of Namaqualand's population were classified as 'coloureds' during apartheid. Between 1798 and 1847, Namaqualand was annexed by the Cape Colony under British rule. This colonial expansion displaced the original Nama population from their traditional grazing areas and caused them to seek refuge at mission stations. These stations were later transformed into so-called 'Coloured Reserves', equivalent of what were known as 'Bantustans' or 'Homelands' in other parts of South Africa.

Since 1994, all of these areas have changed names and are now known as *Communal Areas*. These are areas in which people have shared rights to agricultural and grazing land. However, since these areas primarily served as labour reserves for the mining industry and the white farmers, communal farmers have too little land per person in order to make a living from farming or livestock keeping.

Prior to land reform, the communal areas of Namaqualand represented around 23% of the total area, while 400–450 white farmers owned farms covering 52% of the land. The rest of Namaqualand consisted of mining land (7%), conservation areas (4%), state-owned land (7%) and municipal land (7%). While land reform has led to the redistribution of large farms to communal areas, the poor majority in Namaqualand still lack access to land.

Around one third of the population of Namaqualand are involved in small-scale sheep and goat farming. These people feel connected to the land as part of their identity, and there has been a revival of Nama culture, identity and language since the end of apartheid in 1994. Being Nama is no longer associated with negative concepts such as 'Bushmen', 'Hottentots' or 'Coloureds'. Instead, the Nama have been gaining recognition as an indigenous people (Penn 2004).

We see that the renaissance of Nama culture is closely tied to the question of land. However, land reform and sustainability within Namaqualand's agriculture are also controversial topics debated by scientists, conservationists and activists. This debate is centred on the protection of biodiversity on the one hand and the livelihoods and rights of poor people on the other hand. Natural scientists and conservationists point out that Namaqualand has a unique biological diversity. The area contains more than 3500 plant species, and of these, approximately 25% are endemic—for the most part succulent plants. As a result, Conservation International has classified Namaqualand as one of 25 'biodiversity hotspots' in the world. Scientists and environmentalists have long claimed that the small-scale livestock farming carried out in the communal areas is a threat to biodiversity (Todd and Hoffman 1999; Desmet 2007; Hoffman et al. 2007). In their description of 'overgrazing' and 'desertification' in Namaqualand, Cowling and Pierce (1999:138) wrote for example that

> ... the amount of land under communal use can be expanded – as is, indeed, happening today. Ultimately, however, this land will also be desertified unless some way can be found to manage it within the limits imposed by its production potential.

These attitudes are rooted in the dominant views during colonialism of the agricultural methods of 'native Africans' being ecologically destructive and economically ineffective (Maddox 2002). Today, these attitudes are still central within the dominant discourse of small-scale agriculture in Africa, particularly in South Africa where this discourse is re-produced by the Department of Agriculture, white farmers, national park authorities and other environmental authorities, national and international environmental groups, and many scientists. As a consequence, conservationists and scientists have mobilized funds for expansion of protected areas and implementation of other environmental measures (Hoffman et al. 2007). These funds have come from wealthy South Africans as well as conservation organizations such as WWF and Conservation International.

One may, however, argue that it is ethically problematic to prioritise protection of biodiversity at a maximum level over the fight against poverty in an area that has been through 300 years of repression and marginalization under colonial rule and apartheid (Benjaminsen et al. 2006; Benjaminsen et al. 2008) (see also ► Chap. 7). It seems as if environmental protection has taken over as an argument for a continued uneven distribution of land in Namaqualand. Neither is it in any way certain that the livestock kept by poor people have the negative effects on pastures and biodiversity that many claim (Benjaminsen et al. 2006; Allsopp et al. 2007; Hongslo et al. 2009). After several hundred years of intensive livestock keeping there is still a high endemic biodiversity in Namaqualand. It is also interesting to note that Namaqua National Park was established on some of the hardest used land in Namaqualand. Much of the land that is now part of the national park was previously managed by the mining company De Beers that kept sheep on the land to supply mine workers with meat.

Namaqua National Park was established in January 2002 and thereafter expanded rapidly. In 2006, the park covered 93,000 hectares, and in 2020, it has extended to 146,000 hectares. The park and its rapid expansion have been in direct competition with land reform (Benjaminsen et al. 2008) (◘ Fig. 4.1).

Land redistribution in South Africa is market-based, meaning that the agricultural authorities assist people with little or no land in buying farms that are up for sale. These farms are still mostly owned by white people. Each family has the right to a certain level of support based on certain criteria, which in practice imply that many families (50–100 or more) must come together to manage land, which previously belonged to one white family. When land has been for sale in the areas around Namaqua National Park, the national park authorities (SANParks) have been outbidding the state agricultural authorities. This has meant that instead of the land being redistributed to poor landless people, it has been added to the area of the national park. SANParks has had generous funding available to purchase land, and in addition, white farmers and the mining company De Beers generally prefer to sell land to the national park rather than having it go to redistribution. This seems to be a result of the dominant discourse, which states that small-scale farmers destroy the land they farm.

The first farm that was to become Namaqua National Park was purchased in 1988 by the South African multibillionaire Anton Rupert (1916–2006). The farm was first managed by WWF as a flower sanctuary. Namaqualand is known within South Africa primarily for its beautiful spring flowers. During spring, areas, which some ecologists see as 'disturbed' or 'degraded', are covered by blankets of brightly coloured annual flowers (Photo 4.2). These areas are mostly located along roads, previously farmed land or heavily grazed pastures. It is the spectacular displays of colourful flowers in these places that attract tourists to Namaqualand.

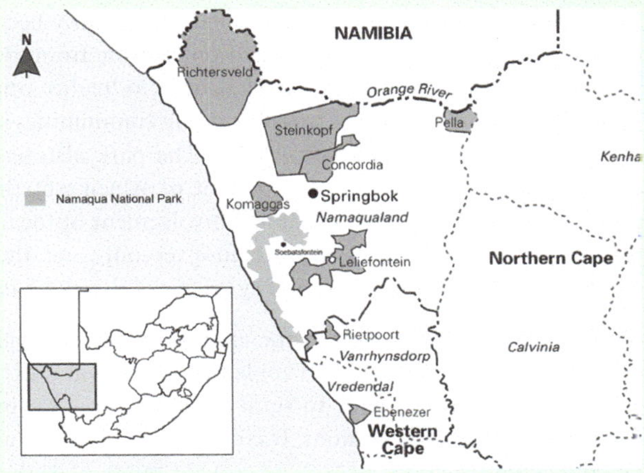

Fig. 4.1 Namaqua National Park. (*Source: Created by the authors*)

Photo 4.2 The annual flowering in Namaqualand. While tourists come to see these displays of flowers in spring (August-September), many ecologists and conservationists see them as signs of 'disturbance' or 'degradation'. (*Photo: Poul Wisborg*)

From the mid-1990s, WWF began, together with SANParks, to plan the establishment of a national park in the area. At the same time, Leslie Hill (1908–2003), another successful South African businessman, established the Leslie Hill Succulent Karoo Trust Fund, which was to finance the purchase of land. This fund is managed by WWF and has been a main source of funding for the expansion of the national park. Both Rupert and Hill have received the WWF recognition known as 'A Gift to the Earth' for their environmental efforts.

Due to the many private wildlife reserves in South Africa, there is an active market for 'wildlife' in the country. In Namaqualand, the park management has purchased various types of antelopes on this market and had for some time also plans to purchase rhinoceros. The original rhinoceros of Namaqualand disappeared around 1880. A high and sturdy fence was built around the park to make it 'rhino-proof'. This fence had two functions: to close in the purchased auction animals and to shut out the resource use of local people. Paradoxically, the landscape in the park, which has a centuries-long history of intensive grazing, is currently classified as 'wilderness' in the IUCN classification system. But still, in 2020, no rhinos have been added to the park.

Even though Namaqua National Park has clearly been established as a fortress conservation practice, the park draws on rhetoric from the win-win discourse. One of the main objectives of the park has been to 'realize opportunities for, and equitable distribution of, benefits to surrounding communities in line with the SANParks resource use policy' (SANParks 2006). The park also has had an 'empowerment' project, the most important component of which was the building of the rhino-proof fence around the park with the involvement of local labour. Thus, people are paid to raise the fence to shut themselves out, and this was curiously labelled 'empowerment'. When we previously critiqued this, SANParks responded:

> » The park's contribution to the empowerment of local people and institutions has been enormous … the People and Conservation Unit of the park continues to actively interact with 16 rural communities, 19 primary schools, three communal farmers associations, two commercial farmers unions, two municipalities and 5 high schools, as well as, a wide range of stakeholder and constituency groups. To date well over 30 000 local people have benefited from environmental education over the past few years. As a result, there is increasing support for the park among local communities as evidenced by demonstrable improvements in the attitudes of local communities towards conservation as a justifiable form of land use. … The Namaqua National Park has a well-established and representative stakeholder Park Forum through which all matters pertaining to park development are discussed. (SANParks in email attachment, October 2005).

Soebatsfontein is one of the local communities near the park. This is a small settlement established in 1954 on land owned by a church, which originally established the settlement to house retired land workers. After retirement, it has been common practice that land workers are not allowed to continue living on the farms where they have worked. In 2005, there were 250-300 residents in Soebatsfontein—both young and old. In the early 2000s, De Beers had begun to close down the diamond mines in the area, and many people in Soebatsfontein lost their jobs. Many therefore hoped that land reform would represent an alternative livelihood. They also felt that they should have prioritized claims to take over the land when neighbouring farms were sold, because many of them had worked on these farms, and some also had their parents or grandparents buried there. Many therefore responded with disbelief when they learned that these farms would be incorporated in the national park. Since then, Soebatsfontein received 15,000 hectares through land reform, but this was far less than their needs. According to the agricultural authorities, this land only had a carrying capacity of 1200 sheep implying that each household could only keep around 40 sheep, while people estimated 150-200 sheep as a minimum amount on which a family could make a living. By comparison, white farmers in Namaqualand have between 500 and several thousand sheep on each farm (Benjaminsen et al. 2008).

To summarize, Namaqua National Park is a case of a fortress conservation practice in a recently established park, despite powerful actors' presentation of it within the framework of the win-win discourse on protected areas.

Tiger Conservation in India

In 2020, India has around 3000 tigers, which is more than a doubling since 2006, and which represent 80% of the global tiger population. To conserve this important species, the Indian government has established 50 tiger reserves that together cover 2.08% of the country's land area. In 2013, more than 51,000 families in 787 villages still lived in the tiger reserves, while about 8000 families had been relocated to other areas. Most people living in the reserves are Adivasis who are indigenous tribal forest dependent communities. They constitute about 8% of India's population (Rai et al. 2019).

The successful conservation of tigers is also a result of a renewed international interest in tiger conservation. The World Bank has for example committed support for tiger conservation by establishing the Global Tiger Initiative, which aims to attract investments into tiger conservation at a large-scale combining tourism with income from carbon credits and other valuations of and payments for ecosystem services. This has been backed by donor commitments at the Tiger Summit held in Russia in 2010 to the tune of 33 million British pounds for countries with tiger populations.

Economic valuation of ecosystem services is a key tool used to support a win-win discourse on tiger conservation and community benefits. In two studies commissioned by the National Tiger Conservation Authority (Verma et al. 2015, 2019) the Total Economic Value of tiger reserves has been calculated. These studies claim to 'make visible' the hidden economic value of conserving forests as important tiger habitats in terms of ecosystem services such as pollination, carbon and water storage, preventing soil erosion, and the protection of genes for potential pharmaceutical developments. In addition to this comes potential income from tourism.

However, Menon and Rai (2017, 2019) point out that these valuation studies assume, following government tiger conservation policy, that forest dwellers will be moved out as part of the conservation measures. Hence, evicted forest people will most likely not be among the beneficiaries. Benefits from the forest will mainly be in terms of pollination and other ecosystem services to farmers close to the reserve as well as wage labour in forest department activities and tourism enterprises. The social benefits from tiger conservation will therefore mostly go to non-forest dependent state and private actors, while the forest dwellers who have used the forest sustainably for centuries are evicted without being among the beneficiaries of conservation. This means that although making visible the hidden economic value of ecosystems sounds intentionally benign, the practical implication is that it serves and justifies exclusion and dispossession.

A management aim of the government has been to relocate people from tiger reserves, but the difficulty in finding available land for relocation as well as resistance to relocation from forest dwellers has resulted in many people continuing to live inside the reserves. As customary livelihood practices have been restricted or banned, forest-based livelihoods have increasingly become difficult.

As discussed in ▶ Chap. 2, displacements caused by conservation practices have been framed in political ecology as a form of accumulation by dispossession following Harvey's (2003) development of Marx's notion of primitive accumulation. According to Li (2009), it is the places and the resources in these places that are valued and not the people. The surplus labour created through dispossession may therefore not to be needed in capitalist production in contrast to Marx's classic example of the enclosures of the English commons, where dispossession of commoners created available labour that could be used in the emerging industrialization in the early nineteenth century.

In many cases, non-capitalist spaces and resources are today opened up for capital accumulation, while local people tend to be in the way of such investments. Hence, they will be 'let die'. This notion is taken from Michel Foucault's examination of 'biopolitics' where sovereign power will either make a population live, through enhancing their health and well-being, or let them die through abandonment (Li 2009). The abandoned people may be seen as a 'surplus population' in the sense that it is surplus to the requirements of capital accumulation. There is limited or no need for their labour in the new economic activities created through their dispossession.

This case study of tiger conservation that we describe here is based on an article by Rai et al. (2019), which examined the consequences of long-term restrictions imposed by conservation on the livelihoods of forest dwellers in the Biligiri Rangaswamy Temple Tiger Reserve (BRT) in the Western Ghats in southern India. The case study demonstrates how the notions of 'accumulation by dispossession', 'letting die' and 'surplus population' may describe the situation of the indigenous Adivasis in this tiger reserve.

The BRT covers 570 km^2 and contains a diversity of vegetation types such as dry scrub forests, woodland savanna, evergreen forests and grasslands (◘ Fig. 4.2). The area was declared a wildlife sanctuary in 1974 and a tiger reserve in 2010.

Fig. 4.2 Map of Biligiri Rangaswamy Temple Tiger Reserve. (*Source: Ashoka Trust for Research in Ecology & the Environment (ATREE), Bangalore*)

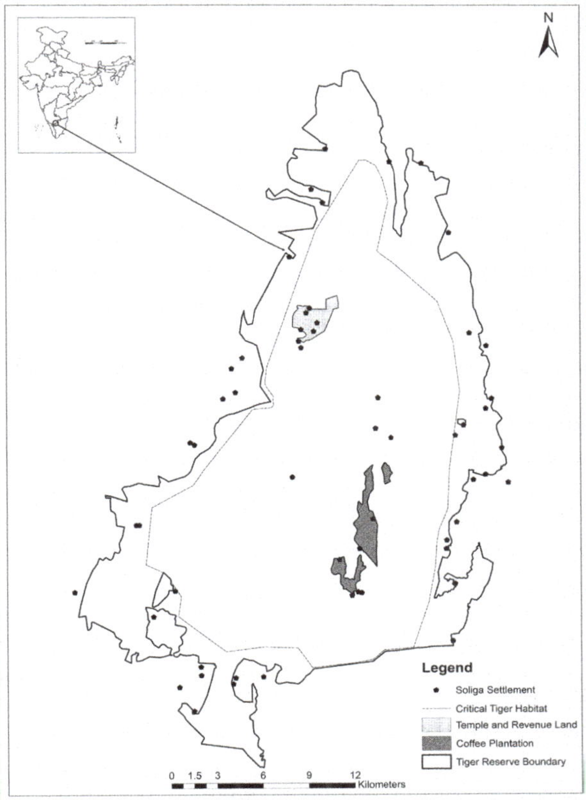

Within the tiger reserve, there are several coffee estates that depend on Adivasi labour. Over 30% of the Adivasi households are employed regularly or seasonally by these estates. Historically, Adivasi hunted small game and assisted in the hunting of larger animals such as elephants and tigers by the local maharajas (kings) and also the British during colonization. The history of the area shows that the landscape has been used and managed by a number of different actors ranging from the colonial state, maharajas, commercial coffee planters, temple authorities and Adivasis. This demonstrates that the area was far from a wilderness landscape that current management strategies strive to recreate.

The Adivasi were shifting cultivators until the area was declared a wildlife sanctuary when they were sedentarized. Prior to the establishment of the sanctuary, Adivasi used to burn the forest to maintain it as savanna woodland with a grassy understory. They cleared areas of forest for the cultivation of millet, beans and a variety of subsistence crops, and they burnt the landscape regularly to maintain an open forest as this ensured visibility, ease of movement and access to a range of products such as tubers and grasses.

From the mid-1970s, hunting was banned in the area. This ban resulted in an increase in wildlife numbers as it has in other parts of India. Herbivores such as sambar (*Rus unicolor*), chital (*Axis axis*) and wild boar (*Sus scrofa*), which were previously hunted for meat, experienced a steady growth in numbers as a result of the decreased hunting. Following this protection and the increase in herbivores, the tiger population in BRT also increased: it is estimated that there were 25 tigers in 2010 and 68 in 2014.

Over time, Adivasi have also experienced increased crop losses caused in particular by the increasing wild boar population. In addition, Adivasi suggest that increased raiding of crops is also caused by the fact that forage has decreased within the sanctuary, due to the increase in density of *Lantana camara*, an ornamental plant turned weed that was introduced to India by the British in the early nineteenth century. Ecological studies have shown that Lantana growth has increased tremendously over the last few decades. There was a three-fold increase in density of Lantana in BRT between 1997 and 2007, and a doubling in the geographic spread within the tiger reserve during this period (Sundaram and Hiremath 2011).

Households that could switch to growing coffee from millet did so to avoid losing crops to wild boars. Farmers switched to coffee as it is not eaten by wild boars and hence considered a safe crop to cultivate. Coffee only grows in wet areas and is not therefore an option for all villages. Moreover, as it takes a few years to produce a yield, only households that have the ability to forego income from agriculture for a few years could afford to switch to coffee.

Adivasi claim that the ban on the use of fire has transformed the once open forest into dense vegetation. Fire kept the undergrowth open and facilitated movement, use and visibility. Lantana densities and cover have increased significantly enough to affect the vegetation dynamics of many tree species due to suppressing growth and recruitment (Ticktin et al. 2012). Adivasi have listed over a hundred species of plants and trees that they say have declined over the past 40 years. This decrease in the population of many plant species as well as the decreased visibility in the forest has resulted in Adivasi not harvesting a range of forest produce. Hence, the combined impact of restrictions, vegetation transformation and wildlife increase has considerably reduced forest-based livelihood opportunities.

Adivasi give an account of forest change that is being confirmed by ecological studies. They describe the impacts of fire suppression as not only increasing Lantana densities, but also that of a mistletoe (*Taxillus tomentosus*) that infects adult trees leading to their early death. The mistletoe is susceptible to the smoke from ground fires and the absence of these fires has resulted in an increase in its density. In corroboration of what the Adivasi have been describing, long-term ecological studies have shown that Lantana has increased significantly (Sundaram and Hiremath 2011); that mistletoe-induced mortality of trees is high (Rist et al. 2008); and that the combined effect of the increase in Lantana growth and mistletoe infection has resulted in the population decline of major trees used for non-timber forest products (NTFP) (Ticktin et al. 2012).

The average annual household income derived from NTFP decreased from 104 US dollars in 1995 to 19 US dollars in 2009 (when corrected for consumer price index). During the same period, the proportion of income from wage labour increased from 37 US dollars in 1995 to 114 US dollars in 2009. The significant reduction in NTFP income and the increase in income from wage labour are consistent with the process of proletarianization that is evident in many parts of rural India (Harriss et al. 2010).

The loss of livelihoods due to the curtailment of access to natural resources has not been compensated by other sources of income. While tourism has increased tremendously during the last decade and provides income for lodges and the state, no portion of these revenues are shared with local people. Neither does tourism offer sufficient livelihood options for people who have been displaced by conservation. The employment of Adivasi in tourism is negligible. In 2010, out of the 31 employees in a state-run tourism resort in the BRT, nine were Adivasi, eight of whom were temporary employees performing low-paid tasks. The 'make live' possibilities that might have existed in the form of employment in tourism or in the redistribution of tourism revenues have been ignored.

Another of the 'make live' options that Adivasi households rely on is now under threat from conservation related restrictions. Households that are located close to the coffee estates have for decades worked in the estates to augment their income. While 17% of Adivasi households in BRT are employed throughout the year in coffee estate work, 13% are employed for a few months when labour requirement is high. Although estate labour is a major source of income for many households, there have been recent attempts by the state forest administration to terminate the leases of the coffee estates. The possible termination of the leases of the coffee estates will have a significant impact on Adivasi and reinforce the 'let die' motivations of state conservation policy.

The alienation of Adivasis from forest-based production and their increased dependence on labour wages and cultivation of cash crops represent a significant transformation in livelihoods as a result of conservation policy. These outcomes have occurred even without the physical eviction of Adivasis from the forest.

The mapping of a web of relations demonstrates how conservation measures have had some primary effects (increase in wildlife density, expansion of Lantana, reduction in NTFP availability) and some secondary effects (increased crop-loss, increased dependence on wages, farmers switching to coffee). These effects further led to a general decrease in total income and in food security for the local population. But as we have seen, restrictions on resource use have also had adverse environmental effects. The ban on fires has resulted in the proliferation of the invasive species *Lantana camara* and reduced forage availability for wildlife due to the reduction of grasses and other vegetation (◘ Photo 4.3).

This case study does not discuss physical displacements caused by conservation, but rather the impact of conservation measures on people who still remain in tiger reserves. Hence, the BRT is a case of in situ displacement (Feldman et al. 2003), which implies socioeconomic, but not spatial displacement (Ince 2014). The case of BRT differs from recently reported cases of conservation as accumulation by dispossession by the fact that the Adivasis are not (as yet) evicted to create 'wilderness' (as is also the case with Wildlife Management Areas in Tanzania discussed earlier).

4

Photo 4.3 Who should be made live and let die? (*Photo: Tor A. Benjaminsen*)

Instead, they maintain settlements within the reserve, but find their land and resource use restricted to the point that they were forced to replace farming with becoming labourers on nearby coffee estates and agricultural lands.

The conversion from independent producer to labourer also involves rapid rural proletarianization (Hall 2011). Adivasi are certainly a 'surplus population' (Li 2009) who are 'in the way' of conservation and accumulation. The chronic squeeze on livelihoods and the resultant proletarianization makes it easy for the state to eventually remove Adivasi from these landscapes.

Conclusions

In this chapter, we have focused on the importance of making an analytical distinction between discourses and practices in conservation. On that basis, we have documented a mismatch between discourse and practice in two cases of protected areas; one in South Africa and one in India. Today the win-win discourse on protected areas dominates presentations on area conservation in Africa, but in both cases, we found practice to be in consistence with the previously dominating fortress conservation discourse.

How can this mismatch between discourse and practice be explained? We argue that both conservation organizations and government officials have interests in reproducing the win-win discourse, albeit for different reasons. It should not be surprising that conservation NGOs are giving priority to conservation aiming at including as large areas as possible under their 'jurisdiction'. They, however, depend on reproducing the win-win discourse in order to attract funding from state aid agencies as well as from companies and wealthy individuals.

State conservation agencies are also important contributors to the win-win discourse on protected areas and are at the same time implementing practices more closely associated with fortress conservation leading to dispossession of a 'surplus population'. In order to attract external funding and tourists to their country, they have an interest in maintaining an impression of a win-win situation between conservation and local people.

Questions

1. Why is there often a gap between conservation practice and discourse in the Global South?
2. Discuss whether, and if so how, conservation practice can become more 'people friendly'?
3. Critics of the radical discourse on conservation argue that its implication will necessarily be environmental degradation and depletion of biodiversity, because of population growth and bad local land-use practices. Discuss these two opposing ideas (the radical discourse vs the revival of the fortress discourse).
4. In what ways can the communities in or around conservation areas be seen as 'surplus populations'?

References

Adams, W. 2004. *Against extinction. The story of conservation*. London: Earthscan.

———. 2009. *Green development*. 3rd ed. London: Routledge.

Adams, W., and D. Hulme. 2001. Conservation & community: Changing narratives, policies & practices in African conservation. In *African wildlife & livelihoods. The promise and performance of community conservation*, ed. D. Hulme and M. Murphree. Oxford and Portsmouth: James Curry Ltd. and Heinemann.

Adams, J.S., and T.O. McShane. 1992. *The myth of wild Africa. Conservation without illusion*. New York: W. W. Norton & Co.

Allsopp, N., C. Laurent, M.C. Debeaudoin, and M.I. Samuels. 2007. Environmental perceptions and practices of livestock keepers on the Namaqualand Commons challenge conventional range and management. *Journal of Arid Environments* 70 (4): 740–754.

Benjaminsen, T.A., and I. Bryceson. 2012. Conservation, green/blue grabbing and accumulation by dispossession in Tanzania. *Journal of Peasant Studies* 39 (2): 335–355.

Benjaminsen, T.A., R.F. Rohde, E. Sjaastad, P. Wisborg, and T. Lebert. 2006. Land reform, range ecology, and carrying capacities in Namaqualand, South Africa. *Annals of the Association of American Geographers* 96 (3): 524–540.

Benjaminsen, T.A., T. Kepe, and S. Bråthen. 2008. Between global interests and local needs: Conservation and land reform in Namaqualand, South Africa. *Africa* 78 (2): 221–244.

Benjaminsen, T.A., M.J. Goldman, M.Y. Minwary, and F.P. Maganga. 2013. Wildlife management in Tanzania: State control, rent seeking and community resistance. *Development and Change* 44 (5): 1087–1109.

Bluwstein, J. 2018. From colonial fortresses to neoliberal landscapes in Northern Tanzania: A biopolitical ecology of wildlife conservation. *Journal of Political Ecology* 25 (1): 144–168.

Bluwstein, J., and J. Friis Lund. 2018. Territoriality by conservation in the Selous–Niassa corridor in Tanzania. *World Development* 101: 453–465.

Bonner, R. 1993. *At the hand of man. Peril and hope for Africa's wildlife*. London: Simon & Schuster.

Brandon, K., K.H. Redford, and S.E. Sanderson, eds. 1998. *Parks in peril: People, politics and protected areas*. Washington, DC: Island Press.

Brechin, S.R., P.R. Wilshusen, and C.L. Fortwangler, eds. 2003. *Contested nature: Promoting international biodiversity with social justice in the twenty-first century*. New York: State University of New York Press.

Brockington, D. 2002. *Fortress conservation: The preservation of the Mkomazi Game Reserve, Tanzania*. Oxford: James Currey.

Brockington, D., R. Duffy, and J. Igoe. 2008. *Nature unbound: Conservation, capitalism and the future of protected areas*. London: Earthscan.

Büscher, B. 2010. Anti-politics as political strategy: Neoliberalism and transfrontier conservation in Southern Africa. *Development and Change* 41 (1): 29–51.

Cavanagh, C., and T.A. Benjaminsen. 2015. Guerrilla agriculture? A biopolitical guide to illicit cultivation within an IUCN Category II protected area. *Journal of Peasant Studies* 42 (3–4): 725–745.

Chapin, M. 2004. A challenge to conservationists. *World Watch Magazine*, November/December, 17–31.

Cooke, B., and U. Kothari, eds. 2001. *Participation: The new tyranny?* London: Zed Books.

Cowling, R., and S. Pierce. 1999. *Namaqualand – A succulent desert*. Vlaeberg: Fernwood Press.

Desmet, P.G. 2007. Namaqualand – A brief overview of the physical and floristic environment. *Journal of Arid Environments* 70 (4): 570–587.

4

Dowie, M. 2005. Conservation refugees: When protecting nature means kicking people out. *Orion Magazine*, November/December 2005.

Dryzek, J.S. 1997. *The politics of the earth. Environmental discourses*. Oxford: Oxford University Press.

Du Toit, J.T., B.H. Walker, and B.M. Campbell. 2004. Conserving tropical nature: Current challenges for ecologists. *Trends in Ecology and Evolution* 19 (1): 12–17.

Dzingirai, V. 2003. The new scramble for the African countryside. *Development and Change* 34 (2): 243–263.

Feldman, S., G. Charles, and S. Louise. 2003. Moving targets: Displacement, impoverishment, and development. *International Social Science Journal* 55: 7–13.

Goldman, M. 2003. Partitioned nature, privileged knowledge: Community-based conservation in Tanzania. *Development and Change* 34 (5): 833–862.

Gramsci, A. 1971. *Selections from the prison notebooks*. Edited by Q. Hoare and G. Smith. New York: International Publishers.

Hall, R. 2011. Land grabbing in southern Africa: The many faces of the investor rush. *Review of African Political Economy* 38: 193–214.

Harriss, J., J. Jeyaranjan, and K. Nagaraj. 2010. Land, labour and caste politics in rural Tamil Nadu in the 20th century: Iruvelpattu (1916-2008). *Economic & Political Weekly* 45 (31): 47–61.

Harvey, D. 2003. *The new imperialism*. Oxford: Oxford University Press.

Heynen, N., J. McCarthy, S. Prudham, and P. Robbins. 2007. *Neoliberal environments: False promises and unnatural consequences*. Oxon and New York: Routledge.

Hoffman, M.T., N. Allsopp, and R.F. Rohde. 2007. Sustainable land use in Namaqualand, South Africa: Key issues in an interdisciplinary debate. *Journal of Arid Environments* 70 (4): 561–846.

Hongslo, E., R.F. Rohde, and M.T. Hoffman. 2009. Landscape change and ecological processes in relation to land-use in Namaqualand, South Africa, 1939 to 2005. *South African Geographical Journal* 91 (1): 4–15.

Hübschle, A.M. 2017. The social economy of rhino poaching: Of economic freedom fighters, professional hunters and marginalized local people. *Current Sociology* 65 (3): 427–447.

Hutton, H., W.M. Adams, and J.C. Murombedzi. 2005. Back to the barriers? Changing narratives in biodiversity conservation. *Forum for Development Studies* 32 (2): 341–370.

Igoe, J., and D. Brockington. 2007. Neoliberal conservation: A brief introduction. *Conservation and Society* 5 (4): 432–449.

Igoe, J., and B. Croucher. 2007. Conservation, commerce, and communities: The story of community-based wildlife management areas in Tanzania's northern tourist circuit. *Conservation and Society* 5 (4): 534–561.

Igoe, J., and S. Sullivan. 2009. *Problematizing neoliberal biodiversity conservation: Displaced and disobedient knowledge*. An Executive Summary for the International Institute for Environment and Development (IIED).

Igoe, J., K. Neves, and D. Brockington. 2010. A spectacular eco-tour around the historic bloc: Theorising the convergence of biodiversity conservation and capitalist expansion. *Antipode* 42 (3): 486–512.

Ince, O.U. 2014. Primitive accumulation, new enclosures, and global land grabs: A theoretical intervention. *Rural Sociology* 79 (1): 104–131.

Kepe, T. 2004. Land restitution and biodiversity conservation in South Africa: The case of Mkambati, Eastern Cape Province. *Canadian Journal of African Studies* 38 (3): 688–704.

Kioko, J., M. Okello, and P. Muruthi. 2006. Elephant numbers and distribution in the Tsavo-Amboseli ecosystem, south-western Kenya. *Pachyderm – Journal of the African Elephant, African Rhino and Asian Rhino Specialist Groups* (40): 60–67.

Kramer, R.A., C. Van Schaik, and J. Johnson, eds. 1997. *Last stand: Protected areas and the defense of tropical biodiversity*. Oxford: Oxford University Press.

Li, T.M. 2009. To make live or die? Rural dispossession and the protection of surplus populations. *Antipode* 41 (1): 66–93.

Maddox, G. 2002. 'Degradation narratives' and 'population time bombs': Myths and realities about African environments. In *South Africa's environmental history. Cases and comparisons*, ed. S. Dovers, R. Edgecombe, and B. Guest, 250–258. Cape Town: David Philip Publishers.

Mariki, S., H. Svarstad, and T.A. Benjaminsen. 2015. Elephants over the cliff: Explaining wildlife killings in Tanzania. *Land Use Policy* 44: 19–30.

Martin, C. 2003. Protected areas: Past, present, and future. *New York Times*, July 29, 2003. http://144.16.65.194/hpg/envis/doc99html/biodpou230830.html.

Menon, A., and N. Rai. 2017. Putting a price on tiger reserves: Creating conservation value or green grabbing? *Economic and Political Weekly* 52 (52): 23–26.

———. 2019. The mismeasure of nature: The political ecology of economic valuation of Tiger Reserves in India. *Journal of Political Ecology* 26: 652–665.

Moss, C.J. 2001. The demography of an African elephant (*Loxodonta africana*) population in Amboseli, Kenya. *Journal of Zoology* 255 (2): 145–156.

Nash, R. 1982. *Wilderness and the American mind*. New Haven: Yale University Press.

Neumann, R.P. 1998. *Imposing wilderness. Struggles over livelihood and nature preservation in Africa*. Berkeley: University of California Press.

Oates, J.F. 1999. *Myth and reality in the rain forest: How conservation strategies are failing in West Africa*. Berkeley: University of California Press.

Penn, N. 2004. Land rights, missionaries and surveyors: Khoisan identity and the Komaggas community. In *History of surveying and land tenure. Surveying and land tenure at the cape 1813–1912. Collected papers. Volume 2*, 151–161. Cape Town: The Institute of Professional Land Surveyors & Geomaticians of the Western Cape.

Rai, N., T.A. Benjaminsen, S. Krishnan, and C. Madegowda. 2019. Political ecology of tiger conservation in India: Adverse effects of banning customary practices in a protected area. *Singapore Journal of Tropical Geography* 40 (1): 124–139.

Rist, L., R.U. Shaanker, E.J. Milner-Gulland, and J. Ghazoul. 2008. Managing mistletoes: The value of local practices for a non-timber forest resource. Forest Ecology and management 255: 1684–1691.

Sanderson, S.E., and K.H. Redford. 2003. Contested relationships between biodiversity conservation and poverty alleviation. *Oryx* 37 (4): 1–2.

SANParks. 2006. Namaqua National Park. *Park Management Plan*, October 2006.

Scott, J.C. 1989. Everyday forms of resistance. *The Copenhagen Journal of Asian Studies* 4: 33–62.

Sklair, L. 2001. *The transnational capitalist class*. Oxford: Blackwell.

Sullivan, S. 2006. The elephant in the room? Problematising 'new' (neoliberal) biodiversity conservation. *Forum for Development Studies* 33 (1): 105–135.

Sundaram, B., and A.J. Hiremath. 2011. *Lantana camara* invasion in a heterogeneous landscape: Patterns of spread and correlation with changes in native vegetation. *Biological Invasions* 14 (16): 1127–1141.

Terborgh, J. 1999. *Requiem for nature*. Washington, DC: Island Press.

Ticktin, T., G. Rengaian, M. Paramesha, and S. Setty. 2012. Disentangling the effects of multiple anthropogenic drivers on the decline of two tropical dry forest trees. *Journal of Applied Ecology* 49: 774–784.

Todd, S., and M.T. Hoffman. 1999. A fence-line contrast reveals effects of heavy livestock grazing on plant diversity and community composition in Namaqualand, South Africa. *Plant Ecology* 142: 169–178.

Veldhuis, M.P., M.E. Ritchie, J.O. Ogutu, T.A. Morrison, C.M. Beale, A.B. Estes, W. Mwakilema, G.O. Ojwang, C.L. Parr, J. Probert, P.W. Wargute, J.G.C. Hopcraft, and H. Olff. 2019. Cross-boundary human impacts compromise the Serengeti-Mara ecosystem. *Science* 363 (6434): 1424–1428.

Verma, M., D. Negandhi, C. Khanna, A. Edgaonkar, A. David, G. Kadekodi, R. Costanza, and R. Singh. 2015. *Economic valuation of tiger reserves in India: A value + approach*. Indian Institute of Forest Management, Bhopal/National Tiger Conservation Authority, New Delhi.

Verma, M., C. Tiwari, S. Anand, A. Edgaonkar, A. David, G. Kadekodi, K.N. Ninan, P. Sharma, P. Panda, and Z. Thatey. 2019. *Economic valuation of tiger reserves in India: Phase II*. Indian Institute of Forest Management, Bhopal/National Tiger Conservation Authority, New Delhi.

Wilshusen, P.R., S.R. Brechin, C.L. Fortwangler, and P.C. West. 2002. Reinventing a square wheel: Critique of a resurgent 'protection paradigm'. *Society and Natural Resources* 15 (1): 17–40.

World Commission on Environment and Development (WCED). 1987. *Our common future*. Oxford: Oxford University Press.

Gender and Power: Feminist Political Ecologies

Contents

T. A. Benjaminsen, H. Svarstad, *Political Ecology*,
https://doi.org/10.1007/978-3-030-56036-2_5

Trailer

On an autumn day in the early 2000s, we were two researchers sitting in the living-room of a farmhouse in Norway. We had arrived to interview a man on this farm, because he represented the landowners in a reference group set up to ensure local representation in policy-making concerning the proposed establishment of a new conservation area. His wife came in to serve us coffee and cake. We invited her to join the conversation, and found that she had interesting thoughts about impacts of the new plans, and that she expressed herself eloquently.

Before we left, we discovered that the wife was the actual owner of the farm and its outfields. She did most of the work on the farm, while her husband held another full-time job in the nearby town. Nonetheless, it was her husband who had been appointed to the local reference group on behalf of the landowners. The establishment of the conservation area was considered to be of great importance to the municipality. The reference group was seen as the principal channel for local views in the process. All actors behind the appointment of the group emphasized that it ought to provide a good representation of the citizens. However, it so happened that not a single woman was present in the group.

Our first reaction was that this case must be a rare coincidence of lack of gender representation in Norway. Nevertheless, it evoked our interest in looking into the status of gender equality in the conservation and environment sector.

Overview

Feminist political ecologies combine elements of feminist thinking and political ecology. In this chapter, we continue the focus on conservation from ▶ Chap. 4. Studies in countries in the Global South might tend to reveal large gender inequalities in policy-making on questions about conservation. In countries in the Global North, however, many take it for granted that women have gained a high degree of equity in policy-making in all fields. Is this so?

Apart from the Norwegian case, in this chapter we also present cases from Senegal, India and Nepal.

5.1 Expectations of Gender Equality in Local Conservation Politics in Norway

There are many reasons to expect Norway to be at the very top of the list of countries with true gender equality in local politics on environmental issues. Some of these are discussed below.

For many years, sustainable development as well as gender equality in politics have been important topics in Norway. Gro Harlem Brundtland received world attention in 1987 when she presented the report from the work of the World Commission on Environment and Development, also called the Brundtland Commission. The report, 'Our Common Future', argued for sustainable development. As the Prime Minister of Norway, Brundtland had also been in world news the previous year, when she appointed a cabinet where eight of the eighteen ministers were women. This became known as 'The Women's Government'. Since then, all Norwegian cabinets have had at least 40% women.

'The Earth Summit'—United Nations Conference on Environment and Development—was held in 1992 in Rio de Janeiro and came in the wake of the Brundtland Commission report. The Convention on Biological Diversity (CBD) was signed at the Earth Summit (United Nations 1992a). Norway was among the countries that played a central role in the negotiations for the CBD, and has since been active in its follow-up. The CBD emphasizes the importance of women taking part at all levels of decision-making in sustainable use and preservation of biodiversity. The Earth Summit also adopted an action plan for sustainable development, known as Agenda 21, which calls for 'Global Action for Women Towards Sustainable and Equitable Development' (United Nations 1992b).

Each year, the United Nations Development Programme publishes a ranking of countries according to the Human Development Index. Norway has always taken up one of the highest positions on this ranking. Gender equality is one of the indicators used, and in this respect, Norway also ranks high, not least due to the relatively high percentage of women in parliament (41% in 2017–2021).

Furthermore, Norwegian policy and legislation make it reasonable to expect a high degree of gender equality. A basic principle is that all public councils, boards and committees shall have at least 40% representation of each gender. This gender balance is required by the Gender Equality Act and the Local Government Act.

During the history of Norwegian aid to developing countries, gender equality has been seen as an important concern. There has been a belief in Norway, based on the country's own achievements in gender equality, could assist countries in the South to improve in a similar way.

In 2004, the Kenyan biology professor and environmental activist Wangari Maathai (1940–2011) received the Nobel Peace Prize in Oslo. One of the key elements of Maathai's work in rural villages in Africa was an emphasis on people shaping their own sustainable development in which women are included as active participants. The Norwegian Nobel Committee awards the Peace Prize, and many Norwegians were proud of its support of sustainable development based on gender equality in Africa.

With such emphasis in Norway on sustainable development as well as on gender, and even on the combination, one would have reason to assume that there is a high degree of gender equality in environmental governance in Norway. After the farm visit described above, we began looking into this.

In our study of local participation in establishing two protected areas, it became apparent that the absence of women among the local participants was one of the most interesting findings. Of 48 representatives in the eight reference groups we studied, only three were women (Svarstad et al. 2006). Thus, we started to wonder whether our experience that day on the farm was perhaps not a coincidence. At least there was a clear pattern of domination by men in the reference groups of the two protected areas we had studied.

When we searched for relevant research knowledge about gender and power in conservation in Norway, we found that such research had scarcely been conducted.

Each year, considerable funds are allocated to research relating to environmental issues. The Research Council of Norway plays a role in the management of much of these funds, while the Directorate for Nature Management and the Ministry of the Environment have a major say in the criteria and allocations of the funds. Over several years, we applied for funding of research on questions of gender equality. Sometimes we experienced our application being put aside without a proper referee process or consideration, because research on gender equality was not considered to be a relevant topic to deal with for the environmental authorities.

At last, we managed to obtain a small grant from the Ministry of Local Government and Regional Development. This gave us the opportunity to further investigate the gendered participation in establishment of new protected areas. We examined the gendered representation in all reference groups set up in connection with the National Park Plan. The implementation of this plan took place from 1992 until 2010 (Photo 5.1).

Forty new and 14 extended protected areas resulted from the comprehensive process, and this constituted major events in Norwegian conservation. We identified 54 comparable reference groups, to which 440 local participants had been appointed. Of these, less than 11% (48 of the 440) were women. In 24 of the 54 groups, there was only one woman, and in 22 groups, there were no women at all.

Photo 5.1 The National Park plan in the 1990s and 2000s consisted of processes to extend the total area of protected areas in Norway, including the extension of Dovrefjell-Sunndalsfjella National Park. This photo is taken at the opening ceremony of the park. From the left: Mayor Erland Løkken, King Harald, and Minister of the Environment Børge Brende. (Photo: Hanne Svarstad)

Only two groups comprised at least 40% women, while the remaining 52 groups did not fulfil the legal requirements on gender equality for such public bodies (Svarstad et al. 2009).

Conservation processes are often regarded as among the most important political issues in rural districts. Much is at stake concerning which areas are to be conserved, and in what ways. Reference groups were established to foster local participation and measures had been taken to provide a good representation of the various local interests and the local population at large, except for the representation of women.

We finally realized that local environmental governance was an area of politics in Norway in which gender equality policies had been marginalized. What could be the reasons? Before discussing possible explanations, we will look at some examples of conservation and gendered power in other countries. In a research project on gender and power in local conservation policies in Europe, we found that Norway is not unique in neglecting gender equality (Klok and van Apeldorn 2007). What about countries in the Global South? We move on to look at research on gendered participation in conservation in Senegal, India and Nepal.

5.2 A Study in Senegal

In a forest area adjacent to Niokolo-Koba National Park in Senegal, under the auspices of the national forest authorities and with support from the World Bank, the creation of a forest reserve took place in 1998–2002 in an area with ten villages. The forest reserve was intended to contribute to conservation and sustainable use of the forest resources. A governance structure was set up for it, based on local management in and between the ten villages.

From early 1998 onwards, emphasis was ostensibly placed on 'gender sensitivity' in this area. Nonetheless, in her doctoral dissertation, Solange Bandiaky documented that women had in fact been given little influence (Bandiaky 2008). The steering committee of the forest reserve comprised representatives from all the villages. Only one of the 14 representatives was a woman. In the village that Bandiaky studied closely, she found that in the committees associated with the forest reserve, there were few women, none of them chairpersons. Through interviews with the women actually appointed to the committees, Bandiaky found that several of them were unaware of their appointment. Accordingly, she concluded that the appointments were fictitious and had been presented in order to satisfy donor requirements of gender equality.

Moreover, she found that women in the positions were not elected by villagers, but named by powerful male leaders, such as their fathers, sons, husbands or friends. The most powerful man in the village was strongly connected with the donors and with the national forest authorities. Villagers who disobeyed him were excluded from essential benefits, such as the distribution of food and seeds financed by the donors (Bandiaky 2008).

To conclude, the research revealed that gendered representation in the Senegalese case hardly existed and did not provide real influence for women. This resembles the situation found in the Norwegian cases.

5.3 A Study in India and Nepal

Bina Agarwal (2001) conducted a comprehensive study of the participation of women in the governance of community forests in five states of India and two districts of Nepal. Commons, such as community forests, are essential for millions of poor people, not least, women who own scarcely any private land. Domestic animals can be grazed and firewood can be gathered in the commons. Over the previous decade, the poor had benefited less from the commons as a result of environmental degradation, state expropriation and privatization. On the other hand, there had been rapid growth in the establishment of local forest groups that govern new commons alone or jointly with the state. However, Agarwal (2001) found that women to a great degree were excluded from participation and influence in these groups. Many of the groups held open meetings in the communities once or twice a year to discuss governance, while boards made decisions between meetings. In the various local forest groups that Agarwal investigated, she found that the average percentage of women members ranged from zero to 10%.

One of the reasons for the low representation of women in several states in India, was found to be that each household could have only one member in the local forest group, and it was expected that the man took this position. There were few women on the boards, and Agarwal found that they often were passive. They were seldom elected to represent other women, and few were aware that they were actually board members.

So, Agarwal's findings of the low representation of women in these cases in India and Nepal are also in line with the findings from the Norwegian and Senegalese cases.

Feminist Political Ecologies

Feminism is a term used in different ways. A central objective is to combat oppression and injustice associated with gender. In ▸ Chap. 1, we presented political ecology as an approach where normative dimensions are explicit. In feminist political ecologies, the normative element implies that studies are intended to contribute knowledge in order to identify and remove gendered oppression and injustice associated with environmental governance.

Dianne Rocheleau is one of the main contributors to feminist political ecology. She has shown how landscapes and ways of living tend to be characterized by various kinds of gendered division of labour and different fields of control over resources by men and women. In 1996, Rocheleau, Thomas-Slayter and Wangari published an edited volume on feminist political ecology (Rocheleau et al. 1996). Its starting point is the focus of political ecology on the decision-making process and the social, political and economic contexts that, in concrete cases, shape environmental policy and practice. The editors contend that political ecology highlights inequitable access to and control of natural resources on the basis of factors such as class, ethnicity, caste, race, age and nationality. Feminist political ecology entails the consideration of gender as an essential factor, together with the aforementioned factors, that are relevant to the access to and control of natural resources (Rocheleau et al. 1996).

Rocheleau has also contributed with a chapter to a more recent book on feminist political ecology (Harcourt and Nelson 2015), where she recounts her experiences with fieldwork over several decades and in different parts of the world (Rocheleau 2015). She presents feminist political ecology as a broad field where different types of feminist perspectives are used. As with political ecology in general, feminist political ecology is characterized by a diversity of approaches and openness to consider inclusions of new ideas from other research fields.

Feminist political ecologies focus traditionally on gendered power and oppression related to land and environmental issues. However, the last years gender tends to be seen in the context of intersectionality in feminist political ecologies (e.g. Sundberg 2017; Sultana 2020) as well as in feminist studies in general. *Intersectionality* may be used as a concept about discrimination against individuals based on mutually constitutive forms of social oppression such as gender, ethnicity, age, class and nationality. Crenshaw (1989) introduced intersectionality into academia through her research on black women and employment in the US (Hopkins 2019).

5.4 From Ecofeminism to Feminist Political Ecologies

In recent decades, scholars, activists and media in many countries have increasingly focused on gender, environment and development, with reference to rural areas in the Global South. The ecofeminist approach of the Indian activist and author, Vandana Shiva, provides a starting point here for a discussion of the substance of feminist political ecologies.

Shiva contends that women and nature tend to be victims of development processes, that women care more for nature than men do, and that women are natural lifesavers of nature (Shiva 1989). Her argument is connected to 'the feminine principle', the source of all life in Hinduism and Hindu philosophy. Shiva depicts the Indian village woman as an embodiment of the feminism principle.

However, colonialism and Western development processes have intruded upon and changed rural villages. Mechanized, market-orientated agriculture has replaced subsistence farming. The change has marginalized small farmers, particularly poor women. Shiva criticizes the Western development model, which she sees as patriarchal and reductionist, and that turns both nature and the work of women into commodities. She believes that this has led to increased wealth in the Global North and greater poverty in the South.

In contrast, she describes traditional Indian agriculture as being in harmony with nature, and with an appreciation of diversity and sustainability. In searching for an alternative global development model, Shiva concludes that the solution is a resumption of subsistence agriculture. It is the poor women of the South who, in their daily struggle to exist, bear the burden of the environmental and development crisis. It is they who know how to survive, so it is they who also have the solution to the problems. According to Shiva, the work of women in subsistence production is sustainable. On the other hand, men take part in the production of goods that degrade the environment. Shiva also contends that women seek development that ensures water and food, while men seek development that generates income (Shiva 1989).

Shiva's ecofeminist theory has gained much attention and is important for the thinking on gender and the environment. The theory brings together perspectives and approaches that previously were separate. It contains elements of feminism and environmental consideration together with dependency theory in which development in the Global North is viewed as based on lack of development or underdevelopment in the South.

For feminist political ecologies, it is relevant to combine some of the same types of elements as those in Shiva's ecofeminism. However, Shiva's theory also contains weaknesses and has been criticized by feminist scholars. A principal objection is that the theory seems to be based on the view that men and women are naturally disposed to distinctive ways of interacting with the environment. Such a viewpoint corresponds to the way feminists of the 1960s and 1970s tended to be concerned with highlighting similarities among women, and contrasting these to similarities among men. This approach has been criticized as constituting *gender dichotomies* with a polarized view of what women and men can and should do in society. Such gender dichotomies associate women with the private sphere and men with the public sphere. Furthermore, women are regarded as being closer to nature than men, while men are associated with intellect and culture. In a later development, feminists instead tend to emphasize great diversities among women, and that they may be situated in various situations which may influence their thoughts and actions differently. Women—as well as men—belong to various classes, castes, age groups and ethnic groups.

Here it is appropriate to look at the difference between sex and gender. Sex designates biological differences between men and women, whereas gender refers to cultural interpretations of these differences. In many areas, there are statistical differences in the choices that women and men make in their lives. The differences may arise from various cultural and social causes, in contrast to the essentialization associated with ideas about biological differences between women and men as the main, or only, causes behind different gender roles.

Anthropologists and sociologists have criticized Shiva's ecofeminism on the basis of empirical research that does not support the general claim of women being more environmentally inclined than men (Jackson 1993; Braidotti et al. 1994). This may be related to a diversity of patterns of gendered divisions of labour in different groups and locations. Often tasks in a society are divided by gender, some being done mostly by men, others mostly by women. However, what is considered to be women's work or men's work varies from society to society and is subject to considerable change over time. Some of this work may cause environmental degradation, others not, and a general gender pattern in environmental effects cannot be concluded.

Shiva criticizes not only the Western development model, but also Western science, which she thinks is based on reductionism. We agree that reductionism can be found in Western science. At the same time, however, her own approach can be criticized for reductionism. Without empirical research, women are depicted as being environmentally friendly in contrast to men, and they are cast in the role of lifesavers of nature. Moreover, women of the Global South tend to be seen as victims of Western development and its resultant environmental degradation. However, social science studies reveal a world that is far more diverse. Shiva's theory nevertheless, it may still be useful to apply as a basis of comparison for students interested in examining specific cases involving gender and the environment.

Shiva's ecofeminism has played an important role in fostering an awareness of the relevance of gender in questions concerning environment and development. In feminist political ecologies today, however, it is pivotal also to learn from weaknesses of ecofeminism. It is especially important to avoid biological essentialism and instead to be aware that gendered divisions of labour, knowledge and power can vary in different societies and cases. Therefore, empirical research has to be conducted in order to establish the various ways that power is gendered.

5.5 Does it Matter That Decisions on the Environment Are Dominated by Men?

We argue that there are grounds for replacing an essentialist and stereotyped perspective of women as being generally more environmentally friendly than men. Does it then matter if men control decision-making processes on conservation and environmental governance? Are there other reasons for ensuring that both women and men participate in political process on these issues?

Our prime argument here is the normative view that women must have the same opportunities as men for taking part in political processes. We argue, in other words, that men and women should have the same rights of democratic participation. We think it should be unnecessary to have to argue for this in Norway or in any other country. The establishment of protected areas and other questions concerning area use are seen as crucial in many communities, and we find it unacceptable to exclude women from participation in policy-making in this field.

Furthermore, systematic exclusion of women as participants in decision-making processes may be regarded as a waste of human resources. Moreover, there is often a gendered division of labour, such that women and men possess different fields of expertise that ought to be taken into consideration. Women and men may also, in particular contexts, have different interests in outcomes from environmental decision-making, owing, for instance, to differences in their ownership and access to land and natural resources.

In individual cases, gendered activities that women are responsible for may be more environmentally friendly than those that men do, but this can never be taken for granted. In concrete cases, researchers may study such gender patterns and investigate causes for the findings. Various factors may be involved, including gendered power and division of labour, cultural notions of women and men, as well as biological differences.

There is reason to point to a paradox in the exclusion of women as local participants in conservation processes. Local participation is a factor that conservation authorities tend to bring in to generate local support of and give legitimacy to a proposed form of nature protection. Excluding women from local participation may make it difficult for governments to attain such goals.

Establishing the Causes of Gender-Specific Power

How can we establish the reasons behind male dominance in decision-making concerning environmental management? In the following, we demonstrate a possible method, that we used in the case from Norway of the lack of gender equality in the implementation of the National Park Plan. We asked how it was possible for the Norwegian government to carry out such a significant process without gender equity. We found it particularly puzzling that this took place in a country so often portrayed as a leader in gender equality (Svarstad et al. 2009).

To provide satisfactory answers, we built on theories of gender and political participation, and we applied an appropriate research approach motivated by Blaikie and Brookfield's chains of explanation and Dorothy Smith's institutional ethnography.

As mentioned in ► Chaps. 1 and 2, Blaikie and Brookfield (1987) suggested that causes of environmental degradation can be studied by first focusing on the immediate environmental managers at the local level. The aim would be to understand the motivations of these managers and why they use land and resources in particular ways. This step may also include a natural science-based study of the environmental impact of the use of the land. Thereafter, the study moves upward to successively higher levels and identifies causes that contribute to explaining the actions of the local natural resource managers.

Dorothy Smith (2005) has devised a similar approach. She is a feminist sociologist concerned with studying the causes of challenges and problems encountered by women in their everyday lives. Smith's institutional ethnography takes its starting point from what she calls people's 'work knowledge' associated with these challenges and problems. She examines such work knowledge and social relationships through interviews on gradually higher levels. With knowledge compiled from all involved levels, the researcher can piece together insight into how the whole structure influences the daily lives of the women that the study began with.

Like Blaikie and Brookfield, we studied questions of power in the management of nature in the Norwegian case described here, although the scope in this particular study was delimited to gendered power. Like Smith, we were interested in the significance of gender as such, although the case study concentrated on political representation. The establishment of new conservation areas involved politics closely associated with the elements of daily work and leisure activities of rural populations. The approaches of Blaikie and Brookfield and Smith both transgress geographical scales through empirical studies of causal relationships by starting at the most local level and thereafter moving via national levels to the global level.

Although we were inspired by these two approaches, we found it suitable to conduct our study of the reference groups in a manner that differed from both of them.

The chain of explanations starts at the most local level and goes upward to an increasingly larger scale. We began asking whether there were so few women in the reference groups because many women had turned down requests to become members. During the research, we often heard the claim that the male dominance of the reference groups was the result of women not being interested in questions of conservation and therefore not being involved in such efforts. If that claim was correct, the study would support a 'supply theory' for political representation. According to this theory, there are few women in politics, because women do not prioritize politics (Norris and Lovenduski 1995).

We asked key actors in the appointments of eight reference groups if anyone they had invited to join a group had declined the invitation, but all the interviewees told us that they had not experienced any such cases. Hence, we concluded that a supply theory could not explain the situation.

Nonetheless, we cannot reject a supply theory as irrelevant in all situations. There may well be a number of reasons why women can be reluctant to accept posts in a field that is otherwise dominated by men. But in order for a woman to reject a post, the post must first have been offered.

By looking into possible causes at progressively higher levels, we found eight causal links that together constitute the chain of explanations for the low participation of local women in the reference groups of the National Park Plan.

Causal link 1: In rural districts, a local elite of men tends to dominate the political arena for management of areas and natural resources. Positions in that arena carry status and provide power on issues considered to be very important. These men are usually well along in years. The Chairmen of Mountain Boards are often former mayors. The Chairman of the Mountain Board in one municipality told us that the mayor of a neighbouring municipality had said that only one person in the municipality outranked him, and that was the Chairman of the Mountain Board.

We have observed a taken-for-granted attitude that local positions on land use and conservation are filled by men. This way of thinking seems also to have been employed in the appointments of local members to reference groups of the National Park Plan.

Causal link 2: Existing councils and organizations concerned with outfield issues were asked to recommend members for the reference groups. These councils and organizations are usually dominated by men, and they recommended few women. Several of the interviewees in the public sector thought that this was a prime reason for the men dominating the reference groups. However, these councils and committees were not asked by the municipalities to recommend both women and men.

Causal link 3: The municipalities recommended reference group candidates. Municipalities are required by law to implement gender equality with at least 40%

representation of each gender in public bodies. However, in the process of the National Park Plan, the municipalities did not take this responsibility, and in 52 of the 54 cases in our study, the gender representation was lower than 40%.

Causal link 4: County governors have the responsibility for carrying through the process for each proposal of a new or enhanced protected area. They are required to ensure that, in such matters, they act in accordance with Norwegian gender equality policy. This responsibility was not taken in any of the 16 counties that we studied.

Causal link 5: The Directorate for Nature Management had the principal responsibility for the implementation of the National Park Plan. In such processes, the Directorate had a responsibility to make sure that all steps were taken in compliance with the Norwegian gender equality policy and laws. However, the Directorate ignored this obligation. The Directorate was also responsible for following up the Biodiversity Convention in Norway, but neglected to make sure that the process followed the Convention in carrying out conservation in full participation of women at all political levels.

When we, in the early 2000s, discussed the situation at the Directorate of their lack of gender equality, we realized that this was not a simple omission, but a deliberate policy, and the Directorate continued to ignore their obligations for gender equality for the rest of the process of implementing the National Park Plan. In addition, proposals to study gender aspects of this implementation were rejected by the Directorate on the grounds that gender was not seen as relevant for environmental governance.

We found the main reason to be related to a broader context in which the process of the National Park Plan was considered demanding, because it evoked conflict lines between local and central governance. The reference groups were introduced as a way to reduce such conflicts. As mentioned above, in many rural communities in Norway, local politics concerning environmental governance and conservation remain dominated by elites of local men. These elites have not voiced a demand for gender equity. On the contrary, we experienced through our research that many men considered gender equality a threat to their own positions of power. Moreover, requests for gender equality were seen as a provocation imposed on rural politicians from the national government. Thus, when the Directorate deliberately avoided meeting their obligations of ensuring gender equality in the process, they avoided provoking the local power elites of men. Consequently, the implementation of the comprehensive process of the National Park Plan could take place without 'unnecessary' resistance.

Causal link 6: Since the early 1990s, Ministers of the Environment have carried the overall responsibility for gender equality within their sector. It was mandatory to establish reference groups in all the National Park Plan processes since 1999. Thereafter, a whole decade passed without a single Minister of the Environment ensuring that the National Park Plan was conducted in accordance with Norwegian gender equality policy and the Biodiversity Convention's emphasis on gender equality. This situation remained, although the Ministry and Ministers of the

Environment both directly and through dissemination of our research results through the media were made aware of the situation from the early 2000s onwards.

Causal link 7: As in many other countries, there has for many years been a so-called integration strategy in place for gender equality in Norway. This strategy obliges all public organizations to conduct their tasks in ways that ensure political and legal equality requirements, and this is supposedly overseen by, in this case, the Ministry of Children and Family Affairs. However, the Ministry failed to realize that the National Park Plan was implemented without requirements of gender equality. In our opinion, this was owing to a basic weakness of the integration strategy itself. The field of local environmental governance has probably been an extreme case of ignorance of gender equality, and, ironically, in such extreme cases where gender equality is ignored by more or less all the involved actors, the situation also becomes invisible to the Ministry in charge of the integration strategy.

Causal link 8: The last chain of explanation is that Norwegian media showed almost no interest in this case of gender inequality. While media usually cover a large diversity of gender issues, the lack of gender equality in questions on local conservation have not been considered important by journalists. The issue mainly affected rural women, and most journalists live in urban areas. The lack of media attention made it possible for all the public bodies and interests who could continue their practices of gender inequality to do so, without being confronted by media with the illegality of these practices.

All in all, our study showed that a strategy based on a responsibility of all public agencies for gender equality in this sector failed. Instead, the result was repudiation of responsibility by all actors at all levels. A vital explanatory factor was that the conservation authorities chose to draw upon the local power elites of men rather than provoke them with requirements for gender equality in local representation (Svarstad et al. 2009).

5.6 Another Day at Work

We started this chapter by recalling a day of data collection in rural Norway, and we now conclude by recalling another such day a couple of years later. We were three researchers attending a meeting of the Dovrefjell Council, which, in the early 2000s, was assigned the responsibility for one of four trial cases for local management of large protected areas in Norway. One of the items on the agenda that day, was that we should present findings from our study of the process to expand the National Park in the Dovre Mountains, as part of the National Park Plan.

One of our principal findings was that the expansion process had taken place only with a modest representation of women in the local reference groups, and that few women had been appointed to the Dovrefjell Council. Representatives were chosen from each municipality and each county authority as well as a non-voting administrative representative from each of them. Since the Dovrefjell Council was regarded as dealing with matters of great importance for the area, the representatives consisted of mayors and other high-profiled figures. Most of the mayors of these municipalities were men. The structure chosen for the Dovrefjell Council made it therefore more difficult to attain gender equality than, for instance, if two political representatives were to be chosen from each municipality (Svarstad et al. 2006).

After our presentation, one of the men of the Council spoke out and proposed that the Council adopted a resolution stating that gender research is not wanted and ought to be stopped. This was an administrative representative from one of the municipalities. Another representative, a mayor, told us that the focus on gender equality was meaningless, because men had been responsible for decision-making on the use of outfields and mountains for thousands of years, and there was no reason to change this. However, the only two local women politicians at the meeting argued that the gender focus was appropriate. Probably in order to avoid public attention, the motion was withdrawn, and the minutes of the meeting did not even mention this discussion. The lack of fulfilment of the legal requirements of gender equality in public conservation boards and councils continued until a Minister of the Environment several years later, in 2009, finally required that the Gender Equality Act should be met also in these fields. However, Lundberg (2018) shows that local boards for conservation management in Norway still tend to have men in leadership positions.

Conclusions

Feminist political ecologies can be mould together through combinations of various elements from feminist theory and political ecology. In this chapter, we have focused on local conservation governance, with presentation of studies from Norway, Senegal, India and Nepal. By looking for explanations at different levels of governance, we have shown how the lack of local gender equity may be a total result of exercises of power at and between different levels.

Power in environmental issues is often gendered. The case study from Norway illustrates a version of intersectionality where women in rural areas are side-lined by local power elites of men together with the national government and media.

Questions

1. How may feminist political ecologies be defined?
2. Do you think that there are reasons to believe that the country where you live (or another country of your choice) has a high or low degree of gender equality in local environmental issues, and what do you think can explain this situation?
3. Search for information about gender representation in a local council or board on an environmental issue, and discuss what the reasons behind the gender (in)equality might be.
4. Do you think gender representation in decision-making on environmental issues matter? Why, or why not?

References

Agarwal, B. 2001. Participatory exclusions, community forestry, and gender: An analysis for South Asia and a conceptual framework. *World Development* 29 (10): 1623–1648.

Bandiaky, S. 2008. Gender inequality in Malidino biodiversity community-based reserve, Senegal: Political parties and the 'village approach'. *Conservation and Society* 6 (1): 62–73.

Blaikie, P., and H. Brookfield, eds. 1987. *Land degradation and society*. London and New York: Methuen.

Braidotti, R., E. Charkiewicz, S. Hausler, and S. Wieringa. 1994. *Women, the environment and sustainable development: Towards a theoretical synthesis*. London: Zed Books.

Crenshaw, K. 1989. Demarginalizing the intersection of race and sex: A black feminist critique of antidiscrimination doctrine, feminist theory and antiracist politics. *University of Chicago Legal Forum* 140: 139–167.

Harcourt, W., and I.L. Nelson, eds. 2015. *Practising feminist political ecologies: Moving beyond the 'green economy'*. London: Zed Books.

Hopkins, P. 2019. Social geography 1: Intersectionality. *Progress in Human Geography* 43 (5): 937–947.

Jackson, C. 1993. Doing what comes naturally? Women and environment in development. *World Development* 21 (12): 1947–1963.

Klok, C., and R. van Apeldorn, eds. 2007. *Gender and biodiversity management and conservation in Europe*. Wageningen: Alterra.

Lundberg, A.K.A. 2018. Gender equality in conservation management: Reproducing or transforming gender differences through local participation? *Society & Natural Resources* 31 (11): 1266–1282.

Norris, P., and L. Lovenduski. 1995. *Political recruitment. Gender, race and class in the British Parliament*. Cambridge: University Press.

Rocheleau, D., B. Thomas-Slayter, and E. Wangari. 1996. Gender and environment: A feminist political ecology perspective. In *Feminist political ecology: Global issues and local experiences*, ed. D. Rocheleau, B. Thomas-Slayter, and E. Wangari. London: Routledge.

Rocheleau, D. 2015. A situated view of feminist political ecology from my networks, roots and territories. In: Harcourt, W. and Nelson, I.L.(eds.). *Practising Feminist Political Ecologies: Moving Beyond the 'Green Economy'*. London: 2ed books.

Shiva, V. 1989. *Staying alive: Women, ecology and development*. Oslo and London: Zed Books.

Smith, D. 2005. *Institutional ethnography. A sociology for people*. Oxford: AltaMira Press.

Sultana, F. 2020. Political ecology 1: From margins to center. *Progress in Human Geography*. https://doi.org/10.1177/0309132520936751.

Sundberg, J. 2017. Feminist political ecology: Sites of inspiration and formation. In *The International encyclopedia of geography*, ed. D. Richardson. Wiley-Blackwell. https://doi.org/10.1002/9781118786352.wbieg0804.

Svarstad, H., K. Daugstad, O.I. Vistad, and I. Guldvik. 2006. New protected areas in Norway: Local participation without gender equality. *Mountain Research and Development* 26 (1): 48–54.

Svarstad, H., Skuland,S., Guldvik, I., and Figari, H. 2009. Fraværet av Likestilling i Lokal naturforvaltning: Nasjonalparkplanen som eksempel. NINA Rapport 432.

United Nations. 1992a. *Convention on biological diversity*. Rio de Janeiro. http://www.cbd.int/convention/convention.shtml.

———. 1992b. Agenda 21. https://sustainabledevelopment.un.org/outcomedocuments/agenda21.

Climate Mitigation Choices: Reducing Deforestation in the Global South Versus Reducing Fossil Fuel Production at Home

Contents

T. A. Benjaminsen, H. Svarstad, *Political Ecology*,
https://doi.org/10.1007/978-3-030-56036-2_6

Trailer

In the documentary film *Before the Flood,* Leonardo DiCaprio travels the world to observe consequences of global climate change. One of the key messages in the film is that without extensive climate change mitigation, everyone will be affected in the times to come, but poor people today will be affected first. DiCaprio therefore urges immediate action to reduce climate emissions. However, he does not point to structural causes of climate change, nor does he seem to significantly reduce his own considerable climate footprint. Instead, he apparently continues an opulent lifestyle, alternating between his many huge mansions, traveling with his own private jet and sometimes relaxing on his yacht, the eleventh largest in the world.

Greta Thunberg, on the other hand, has become world famous for not only speaking about climate change, but also acting accordingly. In 2018, as a 15-year-old high-school student she initiated the school strikes for climate movement. Politicians have talked about climate change for decades, but the young Swede effectively points out how they behave as emperors without clothes. Refusing to board any aeroplane or eat meat, Greta walks the talk when she goes to international meetings and speaks the truth to politicians about lack of sufficient reductions of climate gas emissions. In her speeches about the urgent need for action she relies on the reports from the UN's International Panel on Climate Change (IPCC). (Photo 6.1)

Photo 6.1 After meeting with Greta Thunberg in 2019, Leonardo DiCaprio posted this picture of the two on Instagram noting that he hoped 'that Greta's message is a wake-up call to world leaders everywhere that the time for inaction is over.' (Source: Instagram)

Overview

What are the choices of climate action that are implemented, by which actors, and with what effects and consequences? In this chapter we highlight distributional impacts of some climate mitigation options. It is well-known that insufficient climate mitigation measures today will have negative impacts for future generations. However, it is often not known that the choices of measures also have serious distributional impacts among people who live today. Nevertheless, some governments in the Global North are eager to outsource climate mitigation to countries in the Global South, and forest-based climate mitigation is a common example of this.

In this chapter, we present a case study of such outsourcing from Norway to Tanzania, and we examine consequences as well as causes. We look at climate mitigation choices in terms of what is said and written about these choices in leading discourses, and we compare claims about the various practices with our own studies that examine aspects of the actual practices and reasons why they were chosen for climate mitigation. We also introduce the notion of climate justice, which we discuss in relation to specific climate options.

6.1 The Climate Crisis

Each country has its own history of climate emissions from the industrial revolution to the present. CO_2 is the gas that provides the largest contribution to the greenhouse effect. Historical contributions to greenhouse gas emissions since the mid-nineteenth century accumulate in the atmosphere and cause temperatures to increase. In the last few decades and for the first time in 800,000 years, the CO_2 content in the atmosphere has exceeded 300 parts per million, and the last few years it has even passed 400 parts per million (Ritchie and Roser 2017/2019).

The *carbon budget* reflects the remaining emissions that can be made without causing the global temperature to pass 1.5 or 2 °C. The industrialized nations in the Global North are responsible for enormous historical emissions of climate gases, while China, India and other more recently industrialized countries have modest historical emissions, but huge increases during the last decades. People who live in absolute poverty in either low or medium income countries have, however, scarcely contributed to the climate crisis (◘ Fig. 6.1).

6.2 Choices of Emission Cuts and Climate Justice in Time and Space

Climate justice is a subfield of the normative tradition of environmental justice that we introduced in ► Chap. 2. We argue that 'temporal' and 'spatial' justices are two principal dimensions of climate justice (Svarstad 2021). These two dimensions together provide climate justice in line with the definition by the Brundtland Commission (WCED 1987) of sustainable development as development that meets

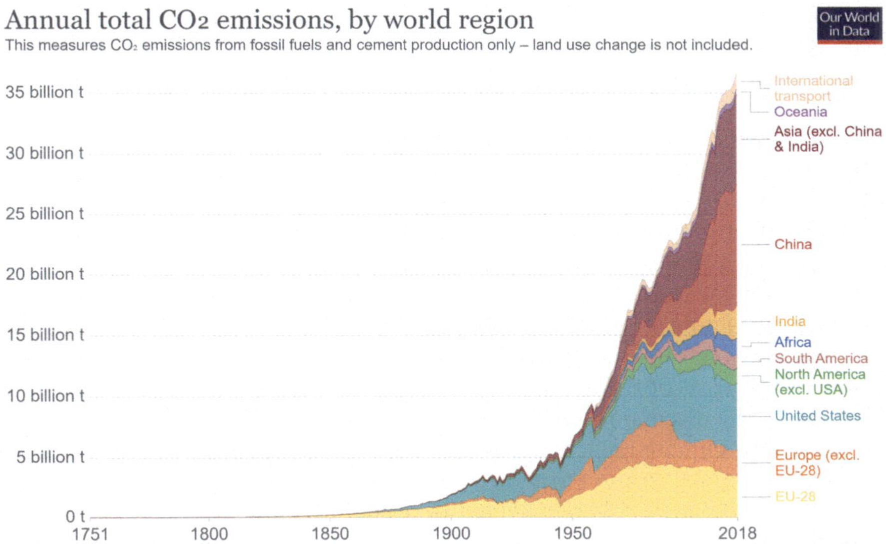

Fig. 6.1 Annual total CO_2 emissions by world region, 1751–2018. (Source: Our World in Data/ Carbon Dioxide Information Analysis Center (CDIAC); Global Carbon Project (GCP)). ▶ OurWorldInData.org/co2-and-other-greenhouse-gas-emissions

the needs of the present without compromising the ability of future generations to meet their needs. Contrary to much of the decision-making about climate change thus far, these are basic principles that would make it possible to establish more democratic processes on climate choices than those which are often seen today. So far, even in countries assumed to have advanced democracies, climate alternatives have often been presented in ways that make it almost impossible for citizens to grasp the most important consequences of each alternative, so that they can make informed choices. Climate justice in time and space are two basic and relatively simple principles that we argue should be fundamental for democratic decision-making on climate mitigation in each country. The principles are similar to what Caney (2014) calls harm avoidance (to avoid harm for future generations) and burden-sharing justice (among countries today).

Temporal climate justice implies that a country's climate policy reduces climate emissions sufficiently and fast enough to avoid the most serious consequences for future generations. Spatial climate justice means that climate mitigation takes place in ways that distribute burden among people today in ways that can be seen as just. This concurs with Caney's principle of burden-sharing justice by putting the responsibility for mitigation on those who have caused the problem, on those who have benefitted from activities causing climate change, as well as on those who have the ability to pay for mitigation. Moreover, as an element of the principle of spatial climate justice, we include the imperative of not placing a mitigation burden on people who are currently vulnerable and who struggle to fulfil their basic needs (Svarstad 2021).

Lately, climate injustice has become an issue within countries in the Global North. The protests against increased fuel costs by the so-called 'yellow vests' in France represent an early example of low-income groups in affluent societies protesting against new taxation, partly initiated for climate mitigation reasons, that have adverse economic effects on these groups.

By the focus in this chapter on climate justice in time and space, we concentrate on the one hand on injustice against people in the future who will be seriously affected if effective mitigation measures are not implemented today. On the other hand, we are also concerned with injustice against people who today are confronted with mitigation measures that deprive them of meeting their basic needs. While the first group of victims of climate injustice is highly present in climate discussions today, the second group is seldom mentioned. In this group, there are people living in remote rural areas of the world, and they are for various reasons ignored in most policy and media discussions of measures for climate mitigation.

Concern for climate justice is to some extent reflected in the international climate change regime. In the Paris Agreement, the Parties to the UN Framework Convention on Climate Change agreed to a goal of limiting global warming to well below 2 °C above pre-industrial levels and also trying to limit it to 1.5 °C. With reference to concern for equity, sustainable development and efforts to eradicate poverty, the Paris Agreement recognizes that developing countries need longer time than other countries to reach the peak in greenhouse gas emissions (United Nations 2015). Thus, the Paris Agreement can be seen to provide a recognition of norms of climate justice between countries, although each country government decides the size of their contributions to emissions cuts. In a special report in 2018, the IPCC argued that in order to limit global warming to 1.5 °C, global human-caused emissions of CO_2 would need to have fallen by about 45% from 2010 levels by 2030 and down to zero around 2050, and this would need much faster reductions than the total promises from the countries in the Paris agreement (IPCC 2018). Greta Thunberg often refers to the urgency expressed in the Paris Agreement and the IPCC's special report on 1.5 °C from 2018, and she emphasizes the need for rich countries to implement immediate cuts and thereby provide opportunities for people in poorer countries to continue to increase their infrastructure (Thunberg 2019).

So, what are the main types of choices for climate mitigation? First of all, there is the choice of cutting emissions from using fossil fuels (coal, oil, gas) as energy sources. For the production of the same amount of energy, of these three sources, coal is responsible for the largest climate emissions, followed by oil and then gas. The use of fossil fuels can be reduced in many ways, such as by restrictions on production and use, fees and taxes, and efforts to change to renewable energy sources such as wind, solar and hydro power. There are also ongoing efforts to develop technology to capture CO_2 emissions as part of the production of fossil fuels and turn it into matter that can be stored permanently, and there are even efforts to capture and store CO_2 that has earlier been released into the atmosphere. Some advocate the need to direct strategies at reducing private consumption, in terms of vacation habits with long flights and changing diets away from meat and food transported over long distances. Some climate action is connected to radical systemic change (such as degrowth). However, governments first of all tend to design their climate mitigation measures in accordance with neoliberal frameworks.

A considerable part of climate mitigation today focuses on forest conservation and afforestation. The idea is to reduce global greenhouse emissions caused by deforestation, since trees absorb and store CO_2 from the air. There is, however, scientific uncertainty about how much impact this may have. In addition, when trees are cut down for energy purposes or when forests burn, CO_2 is again released. Nevertheless, deforestation is a factor contributing to climate emissions, although the estimated size of this contribution has changed.

In 2007, the IPCC's fourth assessment report estimated that deforestation constituted 20% of total climate emissions, but in the fifth assessment report in 2014, this estimate was reduced to about 12%. In addition, Winrock International and Woods Hole Research Center (2013) have concluded that deforestation only represents around 10% of total CO_2 emissions and that this percentage continues to go down, because global climate emissions increase, while deforestation decreases. However, it has become clear that the efforts to arrest deforestation have had limited and often temporary effects of reducing carbon emissions. Over the last years, large forest areas with climate mitigation projects have gone up in smoke in several countries. Brazil is a case showing how vulnerable forests are as carbon storage, when the election of a new president in 2018, Jair Bolsonaro, resulted in a forest policy with focus on exploitation rather than protection.

A premise under the Paris Agreement on climate change from 2015 is that each country makes its own decisions regarding its total contribution to climate mitigation and how to contribute to reduced emissions. Thus, it is up to the country's government whether or not to establish climate mitigation policies in accordance with the principles of temporal and spatial climate justice. In the following, we use Norway as a case for discussing central aspects of a country's climate mitigation policy.

6.3 Norway's Climate Change Mitigation Through Forest Conservation in Tropical Countries

One of us is in a meeting with a group of women in a village in Kondoa District in Tanzania. The women are upset by our question about how they see an ongoing project on conservation of the forest next to the village. They know that we are from Norway, and that the project is funded by Norway. They tell us about how they depend on the forest for daily activities such as collecting firewood for cooking and grazing animals in the dry season. The project has resulted in major restrictions on such activities. One woman says: 'My heart hurts when I hear the name of that project. If you mention it again, I'll leave!'

The project lasted from 2010 to 2014. Since 2011 we have visited the area and conducted fieldwork there many times (Svarstad and Benjaminsen 2017). Each time we have returned to the area, we have met many villagers who have subscribed to a strong narrative of opposition to the project. But we have also met people who have received benefits in the project period, and those who have trusted the promises made by the funding agency and the implementing NGO that the project would provide a reliable source of income for the villagers in the future through sales of carbon credits.

This was one of nine pilot projects financed by the Norwegian government to use forest conservation to mitigate climate change, while also conserving tropical forests. It was part of Norway's International Climate and Forest Initiative (NICFI), and NICFI is an important element of international efforts of Reducing Emissions from Deforestation and forest Degradation (REDD). Proposed at the climate conference in Montreal in 2005, REDD was adopted at the climate conference in Bali 2 years later. A plus sign has been added (REDD+) to indicate additional targets, such as community benefits. However, what is actually added in each case remains unclear, with the plus sign serving a rhetorical purpose rather than reflecting a practice, we therefore prefer to use REDD without the +.

At the climate conference in Bali, Norway's prime minister at the time, Jens Stoltenberg (Labour Party), announced NICFI as a world-leading REDD initiative, and when a conservative government took over in 2013, NICFI continued with the same high level of funding and prioritization within Norway's climate policy. At the Bali conference, Stoltenberg promised an annual funding of 3 billion NOK (about 300 million US dollars). By the end of 2017, Norway had in total spent 22.4 billion NOK (2.24 billion US dollars) on NICFI (Government of Norway 2018) (◘ Photo 6.2).

◘ **Photo 6.2** September 2008: Brazilian and Norwegian leaders meet to sign an agreement for protection of tropical rainforest as part of the Norwegian REDD programme NICFI. From the left: The Brazilian Minister of the Environment, Carlos Minc, Norwegian Prime Minister Jens Stoltenberg, Norwegian Minister for Environment and Development Erik Solheim. (Photo: NICFI)

6.4 The Case of a REDD Project in Tanzania

In 2008, Prime Minister Stoltenberg and the Minister for Environment and Development, Erik Solheim, visited Tanzania and announced that Norway had granted 500 million NOK (50 million US dollars) over the next 5 years to REDD activities in the country. This decision was surprising, since an evaluation report and a detailed auditing in 2006–2007 had revealed extensive corruption and non-fulfilment of requirements of Norwegian environmental assistance to Tanzania during 1994–2006 (Jansen 2009, 2014).

Under this new climate funding to Tanzania, nine pilot projects were facilitated by international and national NGOs. The project in Kondoa that we studied, was implemented by the African Wildlife Foundation (AWF). At the start of the project, there were 21 villages on the hill ridge that constituted the project area where about 62,000 inhabitants were living of whom 28% were estimated to live in deep poverty with an income of less than one dollar a day. According to assessments by Mung'ong'o et al. (2011), a large part of the rest of the population can also be considered to be poor. Nineteen of the villages participated in the project, while the village assemblies in two villages decided not to take part, although they also experienced impacts of increased restrictions on forest resources. In addition to the villages adjacent to two large forest reserves, some other villages participated in the project due to their possession of smaller village forests (Fig. 6.2).

According to the project proposal and the contract with the Norwegian Ministry of Foreign Affairs, the goal of the REDD project in Kondoa was to contribute to poverty reduction and climate change mitigation, and the purpose was to prepare for participation in carbon markets.

The main focus in the project was to implement a strict regime of conservation of the forests of the Kondoa-Irangi Hills as a measure to mitigate climate change. Restrictions on land use concerned activities such as grazing, collection of dry branches for firewood used for cooking, cutting of trees for timber and production of charcoal. AWF chose this particular hill ridge because it constitutes the water catchment area for the river Tarangire, which is the water source for the abundant wildlife in and around Tarangire National Park. For AWF, the REDD pilot project, through establishing a healthy and well managed forest, was seen as an opportunity to secure stable water supplies during the whole year for the wildlife down on the plain below the forest.

In order to compensate people for their immediate losses of access to forest resources and instead contribute to the goal of poverty reduction, the project aimed to provide livelihood alternatives. An agricultural component was from an early stage of the project presented as particularly successful and a key to the overall success of the project. Other project attempts to strengthen livelihood activities might be positive in themselves, but the implementation and associated benefits turned out to be limited. This was the case with tree planting, sustainable production of bricks and charcoal, as well as the introduction of more energy-effective cooking stoves.

As a long-term income source for the villages in the project area, AWF tried to obtain certification for selling carbon credits in carbon markets (see the presentation later in this chapter about carbon credits and carbon markets). People were

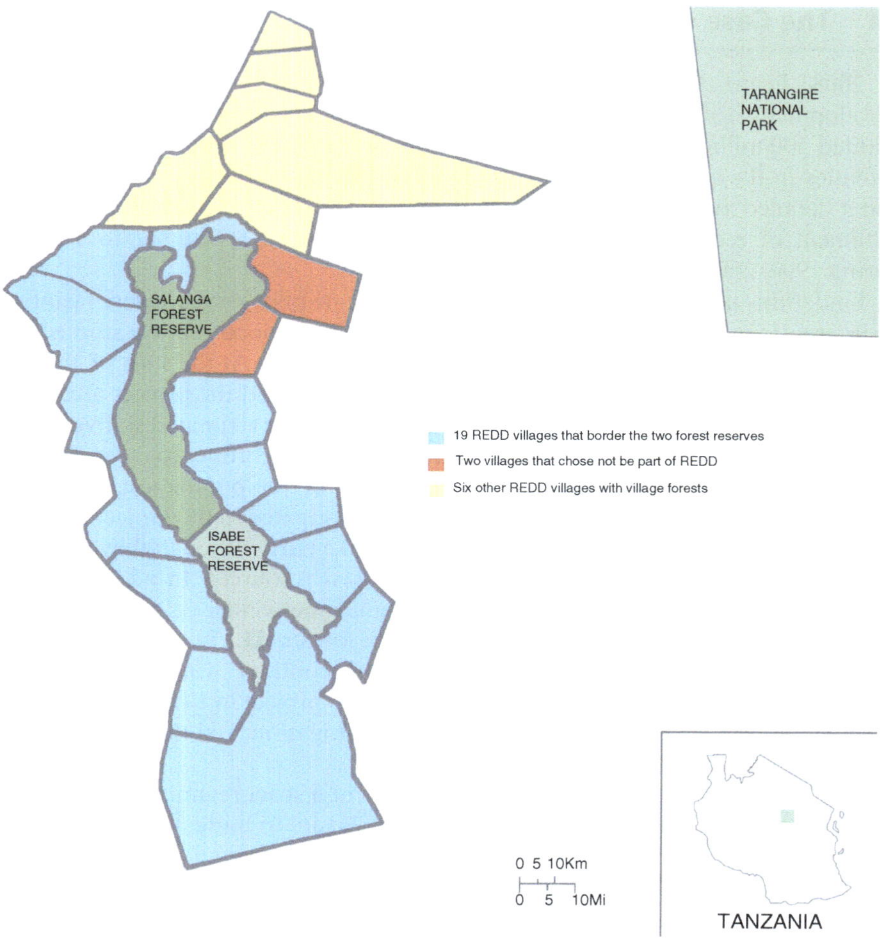

6

Fig. 6.2 Map of the area of the REDD project in Kondoa, Tanzania. (Source: Created by the authors)

told that if their village joined the project, they could expect substantial income in the future through carbon sales on these markets. For several years, AWF reported that this element of the project was in progress, but eventually it became clear that they did not manage to fulfil this promise.

AWF facilitated the establishment of a new organization to manage the forest with membership from the various villages bordering the two forest reserves. From each village, four villagers were recruited as forest guards who received uniforms, some training, and were required to take part in forest patrolling. Villagers caught for illegal trespassing and activities in the forest were given stiff fines, and some were jailed. Permits were, however, granted for some resource use such as collection of dry firewood, but with limitations on when and how often collection could take place and how much firewood could be collected each time. Many women collecting firewood considered the permit fees too expensive, and they chose instead to walk longer distances to collect wood elsewhere.

We found that three overlapping groups in particular were adversely affected by the restrictions on forest use. First, people of the 21 villages living close to the conserved forests and without alternative forested areas nearby tended to be more seriously affected than others. Second, villagers with relatively small farms or without farmland at all were more affected than others. This is because many villagers who lack sufficient farmland depend more on forest resources to sustain a living, for instance, by charcoal production. Third, women tended to be more affected than men, because of their roles in the gendered division of labour, and particularly with collecting dry firewood for domestic purposes (Svarstad and Benjaminsen 2017). (Photo 6.3)

The essence of the agricultural component was that 12 farmers in each of the 19 project villages were appointed as 'demonstration farmers'. They received training in 'improved farming methods' and agreed to make one acre of their farmland available as demonstration plots. On these plots, they were to apply the methods learned, and a central element was to use improved seeds including hybrids, chemical fertilizers and pesticides. Furthermore, farmers were taught to change from intercropping to the alternation of two crops in every second row, and to use a rope to align plants in straight rows and with regular space in-between. On hill slopes, farmers were advised to place the rows along con-

Photo 6.3 Restricting local forest use had negative consequences for women who used to collect dry firewood for cooking. (Photo: Hanne Svarstad)

tours to reduce erosion during heavy rains. The intention was that other villagers should adapt this approach by following the examples of these demonstration farmers.

AWF used the label 'conservation agriculture' for this farming approach. It is, however, more in line with mainstream agricultural modernization primarily aiming at increasing yields through adding external inputs. AWF introduced this approach as a way to reduce the pressure on the forests and to increase agricultural productivity, in order to reduce forest dependence. This differs substantially from FAO's established version of conservation agriculture aimed at more environmentally friendly conditions on the fields through reducing soil disturbance, maintaining soil cover and practising crop rotation.

The maize hybrids promoted by the REDD project came from Pannar and DuPont Pioneer (both owned by the American corporation DuPont), while the open pollinated varieties are bred and produced by Tanzanian research stations. The agricultural component of this project was to a large extent consistent with the green revolution approach that has been tried for several decades in Tanzania and still dominates within the state and among aid donors.

As with the other project components, the agricultural component was characterized by weak administration, too few extension workers, and too little funding to carry out the project according to the aims. There was also scepticism among the villagers, because the introduced package would increase their vulnerability. This is a relatively dry area with unpredictable rainfall leading to frequent crop failure. With the project strategy of modernization, smallholders would in dry years not only lose their crop, but also lose the funds invested to buy seeds, pesticides and fertilizers. They therefore risked being seriously indebted.

In political ecology, there is a considerable scholarship demonstrating how local communities in rural Africa may resist top-down conservation efforts (e.g. Cavanagh and Benjaminsen 2015). In Kondoa, we have seen how resistance from the very beginning has been exercised in terms of continuation of the use of forest resources in ways that had become prohibited. Kondoa actually has a history of restrictions on local resource use going back to colonial times. The two forest reserves were established in 1941 and 1954, and restrictions on forest use have varied considerably since then. Some forest use is harmless, such as collection of dry firewood. Other uses may have more serious consequences, including large-scale logging for building materials and charcoal production. Resistance to conservation in this area sometimes takes place by setting the forest on fire in one part of it, and when forest guards arrive to extinguish the fire, a new fire could be set in another part of the forest. Violent attacks on forest guards have also taken place. At the end of 2015, an election for a new district council was held, and the forest governance that had been established by the project became a central issue. Most representatives who were elected were critical of the REDD project. During later visits, we have observed that, to a large extent, the forest management, which the project established, has stopped working. An explicit intention of the REDD project was participation by the villagers in forest management, but this was never really implemented. Instead, the national government has taken over forest management through its Tanzania Forest Services Agency.

A satellite image study has been made of forest change in the Kondoa-Irangi Hills and of four other sites of the REDD pilot projects, including in nearby control areas, over the period 2008–2018. No significant forest changes was found (Munsterhjelm 2019). Hence, this REDD project seems not to have achieved its aims of climate mitigation nor poverty reduction. Nevertheless, the main actors behind the project have presented the project as a particularly successful case, with an emphasis on the agricultural component. In 2012, the project was highlighted in several pages in a booklet about ongoing Norwegian development assistance to Tanzania (Royal Norwegian Embassy in Dar es Salaam 2012). Under the heading 'Increasing food production to reduce emissions', the embassy claimed that the training of smallholding farmers resulted in an eight-fold increase of maize production without increasing the farmland. Later, on their webpage, the embassy asserted that the project had resulted in food security and economic improvements (Royal Norwegian Embassy in Dar es Salaam 2014).

Both the Norwegian Embassy and AWF have repeated these claims in written presentations and in interviews with us. Large increases in maize yields in particular have been claimed, although the claims have ranged from three to eight-fold increases. AWF also claimed that 1600 farmers annually adopted the agricultural approach that they introduced with the project.

This win-win narrative has striking similarities to other win-win narratives about environment and development that we present in other chapters. The producers of this win-win narrative consist primarily of the actors behind the project, which are AWF and Norway. The message is that the whole globe benefits from the project's contribution to reducing global warming, and that people in the area benefit through poverty reduction. This narrative about a specific project can be seen as feeding into, as well as possibly drawing from, the win-win discourse about REDD in general.

We found that there is a lack of documentation to support the claims of the large increases in yields and the widespread transition to new farming methods. When we asked AWF about such documentation, they told us to ask the Norwegian embassy, and when we asked the embassy, they said we should ask AWF. The claims of six- or eight-fold increases of yields is far from realistic. In an area as dry as this, even a tripling of yields would be impossible without irrigation.

There were many problems with the implementation of the agricultural component of this project, such as poor organization, and limited training and follow-up of the farmers. Input factors were delivered too late to be available in the beginning of the growth season, and there was disappointment among the demonstration farmers because they received support for only one agricultural season.

Moreover, only parts of what was presented as results of REDD could actually be attributed to this project, because similar support to agricultural modernization in the same area and the same period also came from other sources, in particular the EU, the US and the Tanzanian government. The overlapping funding makes it impossible to isolate the effects of the REDD project. Besides, the funding by the three donors has only had limited effects on agricultural modernization overall.

A study of the level of adoption in the REDD project area found little evidence that the agricultural component had a significant effect on rural livelihoods and farming methods. The results showed modest adoption rates of practices promoted by the REDD project. For instance, the rate of agro-chemical use in one of the project villages, which was frequently used by AWF as a demonstration village when receiving visitors, was found to be the same as in villages that did not participate in the REDD project (with 6% of farmers using synthetic fertilizers and 18% using pesticides) (Putnina 2019).

We found, in other words, that AWF and the Norwegian embassy could not substantiate their claims of the success of the REDD project with high yield increases and diffusion of new technology. Likewise, there was a lack of substantiation that agricultural modernization and the other livelihood elements had caused poverty reduction or even compensated the costs that the closure of the forest implied for local people.

6.5 How Can a Success Narrative About a Failed Project Succeed?

As shown in several chapters in this book, there is often a gap between leading discourses and what researchers find when they carry out in-depth studies of cases. So, how is it possible that a failed project can be presented as a great success? And why is the failure not revealed—unless, as in this case, there happens to be an opportunity to carry out independent and in-depth research?

An analysis of these questions has to examine both actors and structures. First, we have actors who, for different reasons, may present the project as a success in order to serve their own interests. According to Chapin (2004) and Büscher (2014), 'success' in conservation and development aid relies upon successful marketing. Chapin discusses, for instance, how big conservation NGOs market their activities in ways similar to those of multinational corporations.

It is worth noting that individuals as well as their organizations may have interests in establishing a success narrative. An organization that manages to present its projects as successful, will make use of this success in new fund-raising. For individuals in these organizations, success narratives contribute to securing their jobs and increasing their competitive strength in the labour market. This, in other words, illustrates the saying: 'nothing succeeds like success'.

In a similar vein, it is important for developing aid administrations and their employees to present the activities that they fund as successful. Furthermore, there is 'the pipeline problem', the need to spend each year's budget before the end of the year (Jansen 2014). Negative attention can lead to delays and be damaging for the organization and programme officer.

Second, some villagers and village leaders seem also to have interpreted the project and marketing of its success as being in their own interest, while others have taken contrary positions. When the project has been presented by AWF and the embassy, some of the actors who have benefited get to play their roles in order to communicate a glossy picture of success.

Third, structural frames tend to prevent interventions for development or conservation purposes to be subjected to evaluations and research that are independent and comprehensive enough to uncover actual impacts. Consultants who have evaluated the pilot projects in Tanzania, for instance, have had very limited time frames and resources. Like the embassy personnel, they have either not visited the project areas at all, or they have been there a day or two under the guidance of AWF, and been taken around to local people and villages selected by AWF. Jansen (2014) mentions that during his long career in development aid, he observed how consultants who have provided critical observations are omitted when new consultancy work has been announced. He also observed that embassies and aid administrations sometimes shape mandates to prevent critical consultants and researchers from being formally qualified.

Fourth, we should also look at the role of the media. In Norway, the media has hardly investigated what happens in practice when the country has taken the international lead in REDD. There have been remarkably few examples of journalists taking a critical look at this leading climate mitigation programme.

Could the reason for the lack of critical and independent media attention be that its social consequences represent what Rob Nixon (2011) calls 'slow violence', which implies 'violence' on livelihoods that takes place over time and which is therefore not recognized as news?

We argue instead that a major reason for the Norwegian media neglecting REDD has been that the REDD events take place in remote places and far away from donor countries such as Norway. For journalists it would be time-consuming and difficult to understand how REDD in a particular area influences the situation of villagers. Hence, since the social consequences of REDD are largely invisible to the citizens in the donor countries, these impacts can be an example of commodity fetishism (explained in ► Chap. 2).

Fifth, the win-win discourse on REDD is produced by a group of actors who have interests in the discourse and related practices becoming a new fad with promises of positive changes. This might, in different ways, be the case for organizations, consultants, government employees, journalists and politicians. REDD may be seen as one of the last in a row of such environment and development fads (Lund et al. 2017). In the case of REDD, the core concern is climate mitigation based on forest conservation, but we think the same phenomenon also occurs in other fields of development assistance. Each fad tends to start with the introduction at an international event, before programmes and projects are planned, carried out and eventually replaced with new fads. After each fad, their modes of management tend to fall apart (as we observed in the REDD case in Kondoa), while some management elements may also become more permanent.

Sixth and finally, there is a reason to ask why politicians, in cases of big spending on international climate strategies, have not requested independent studies of impacts of this enormous funding in the many countries and localities around the world. Political parties in the Norwegian parliament have not shown any interest in knowing how this climate mitigation programme is carried out in practice.

In February 2015, a delegation from the Norwegian parliament visited Dar es Salaam, and the embassy organized a meeting about the experiences with the Norwegian REDD programme in Tanzania. Pilot projects were presented by the organizations who had been given the responsibility for carrying out these projects. One of us who was in Tanzania at the time, requested the embassy's permission to observe this meeting, but the request was turned down. In this way, the success narrative could be presented to Norwegian parliamentarians without the risk of them learning about research which concludes in contradiction to the success narrative promoted by the embassy.

Politicians are, however, not just victims in this case. They should have made sure that there is a system for reliable knowledge production about the consequences of policy-making. In the absence of independent knowledge production, it is possible for success stories about failed projects, programmes and policies to thrive.

Methods: The Relevance of Case Studies

A case study can be defined as an in-depth study of a single example of a category of phenomena. Most political ecologists carry out case studies. Can such studies generate knowledge of value beyond the specific cases? Kondoa is but one among many areas around the world with REDD interventions. Can such cases provide insights of value about REDD in general? Or is the value of a single case study always limited to the case itself?

In 2018, we communicated research findings on REDD through an opinion piece in the largest newspaper in Norway. We told about how the project in Kondoa was presented by the Norwegian government as an illustration of the success of their REDD programme, although our research findings did not confirm this success. The Minister of Climate and Environment at the time, Ola Elvestuen (Liberal Party), responded and dismissed our critique because it was based on a single case study. In public debates like this, we have learned that many people believe that case studies bring 'anecdotes' rather than valuable knowledge.

The case study specialist Bent Flyvbjerg argues that there are crucial knowledge gains from case studies other than generalization, and that the claim that one cannot generalize from one case is also wrong (Flyvbjerg 2006). He refers to the example Popper (1959/2002) used about falsification: With the proposition that all swans are white, only one case of a black swan would falsify this.

In line with Flyvbjerg, we also argue that knowledge from case studies provides elements that are crucial in order to understand larger topics. We will mention three ways that case studies can bring important knowledge about a broader theme, such as the local consequences of REDD.

First, there might be special features about a particular case that makes an in-depth study more valuable than studies of other cases. As shown both in this and in other chapters, we think that cases with strong narrative claims, for instance, of success or of failure, make in-depth insights particularly valuable. This is even more valuable when the claims are presented in a discursive narrative of a leading and thereby powerful discourse. In the Kondoa case, Norwegian government officials widely disseminated presentations of this case as an example of how successful

REDD interventions are. Hence, according to the main funder of REDD, this project was particularly successful. When in-depth research of this case reveals that the success claims are based on weak evidence, it is a finding that sheds a critical light on the REDD programme in general and how it is 'sold' to a wider audience.

In the case study literature, one might find recommendations for choosing cases based on previous assumptions that they are, for example, 'extreme' in the sense of being 'typical' or 'crucial' in relation to a particular phenomenon: 'If this is (not) valid for this case, then it applies to all (no) cases' (Flyvbjerg 2006). Such case selections might sometimes be possible to do when there are strong and dominating claims about a particular case. But often this is knowledge that a researcher only gains after having worked for a while with a study. For instance, we selected and started the case study in Kondoa before it was presented as a success case, and it was only later when it was claimed by the Norwegian government to be a particularly successful case of REDD, that we could consider our study to be 'a crucial case' (sensu Eckstein 1975).

Second, case studies may provide knowledge about *mechanisms* that also might be relevant to other cases. In the REDD case in Kondoa we found that the strict closure of the forest tended to have more serious consequences for the poorest people, for those who had little or no access to farmland, and for women. Collecting forest resources was more important for these groups than for other villagers, they had less access to other livelihood alternatives and the compensation for the loss of forest livelihoods required resources that the poorest did not have. Such insights require in-depth studies using qualitative methods. When doing other case studies, it would be relevant to investigate whether or not the same—or similar—mechanisms are in place.

Third, case studies on a topic contribute to *cumulative knowledge*. Researchers who have conducted an independent in-depth study of a single case are also expected to stay informed regarding research knowledge about other cases on the same topic. In-depth knowledge from one or a few cases makes it possible to have broader comparisons across sites and countries.

When the phenomenon in question is an extensive activity of interventions in remote areas of the Global South, we find it important to have as many in-depth case studies as possible conducted by researchers who are independent of the actors behind the interventions. In the absence of such in-depth studies, the possibilities for understanding the impacts of REDD in different localities around the world would be limited.

In this chapter, we also write about Norway as a case country in the Global North that during the last decades has elaborated a national climate mitigation policy, and we discuss how and why the Norwegian REDD programme has been established as an important element of this policy. We argue that one of the reasons this knowledge is important, is that it constitutes part of the cumulative knowledge about REDD interventions. What happen on REDD sites around the world are not merely the results of local or even national contexts. These interventions must be understood in terms of how they also have emerged as responses to contexts in the Global North.

6.6 The Win-Win Discourse on REDD: From Hegemony to Challenges

The main message that is presented by those who contribute to the win-win discourse on REDD is that actors in the Global North should pay for efforts to mitigate global warming by conserving forests in the Global South. In Norway, we have seen how this has been presented as a cost-effective method of climate action in comparison to climate cuts in Norway. It is also argued that the resulting forest conservation contributes to conserve biodiversity and to provide new sources of income to forest adjacent communities. REDD proponents also often refer to 'safeguards' that the UN have established to avoid negative consequences of REDD for vulnerable people (UN-REDD Programme 2012). Within the win-win discourse on REDD, there is discussion about whether the payment for carbon storage in forests should primarily be made through market solutions based on carbon credits, or according to agreements between funding and forest-owning governments.

Some of the most eager contributors to the win-win discourse, are actors who are among the main funders of REDD programmes. These are the governments in countries such as Norway, Germany and the UK. Likewise, there are governments who have received or hope to receive REDD funding. In addition, international organizations such as the World Bank and the UNDP also contribute to the production of the win-win discourse. Moreover, large international conservation organizations have their own interests in REDD and also adhere to the win-win discourse on REDD. In the case of Tanzania, these include WWF and AWF.

In Norway, the win-win REDD discourse is hegemonic. This implies that it has hardly been contested in public and political debates, and the policy about REDD is almost totally based on its premises. Since 2007, most parties represented in the parliament have taken part in cabinets in which REDD has constituted a key element of the government's climate policy. So far, only the right-wing progress party has a policy that is critical towards REDD.

As we have seen in previous chapters, narratives about specific success cases often play important roles within win-win discourses on environment and development. On the contrary, the counternarrative that we present on the basis of our findings from the case study in Kondoa is in line with—and thereby contributes to—a radical REDD discourse which has clear similarities with other discourses that we labelled 'radical' in ► Chap. 3.

6.7 Three Types of Carbon Trade

What is carbon trade? The owners of a plantation in Uganda may be paid for tree planting through carbon trade. Some of that payment might come from a family on their way from the US to a vacation in South Africa. They pay a voluntary fee to a carbon trade company and are told that their trip therefore is 'carbon neutral'. An energy company in the Netherlands may buy 'carbon credits' in order to be permitted to continue to burn coal with large emissions of CO_2. The money might go to a company that has established a large windmill park in Mexico.

From one perspective, carbon trade is seen as a way of reducing global warming that, at the same time, is beneficial to all involved actors. From another perspective, carbon trade is considered a way to justify continuous carbon emissions from wealthy actors, and with negative impacts on temporal as well as spatial climate justice.

Carbon trade is based on the establishing of new types of markets where buyers and sellers can link up through specialist brokers. Governments specify legal conditions for carbon trade, and consultancy companies are involved in validations of various aspects related to the trade. Either the buyers pay to get formal rights to emit climate gases, or they buy carbon credits on a voluntary market to make up for their carbon emissions. Sellers, on the other hand, receive payments to carry out efforts to reduce climate emissions. The 'commodity' is measured in tons of CO_2 equivalents. As a simplification, some call the emission 'carbon' and the trade 'carbon trade'. A document must be made specifying how many tons of CO_2 equivalents are involved in a particular trade. One ton of CO_2 equivalents involved in such a trade is called a carbon credit.

Since climate change entered the political agenda in Norway in the beginning of the 1990s, governments on both the left and the right side have established and implemented a climate policy that places a substantial portion of climate cuts outside Norway's borders. So far in this chapter, we have first of all focused on REDD with its forest conservation in tropical countries as a key element of this approach. Carbon trade is a different element, but it is also sometimes connected to REDD. Both REDD and the carbon trade have been presented as cost-effective for Norway. Furthermore, an important intention by some of the actors behind REDD has been that carbon trade will finance REDD activities.

One may distinguish between three types of carbon trade. First, there are voluntary markets for carbon credits. In the REDD project in Kondoa, the stated purpose was to prepare for participation in carbon markets, and AWF aimed for voluntary markets. The intention was that the emission reductions should create local income, and thereby secure continued forest conservation after the end of the project. However, AWF did not manage to obtain a required certification. Such certification is a process led by specialist companies. The process may be long and expensive, and many criteria have to be met.

Buyers in voluntary markets consist of companies, governments and private individuals such as business travellers or tourists on holiday. They all want to 'offset' carbon, implying that they see their purchase of carbon credits as funding the mitigation of their own contributions to global warming. For instance, a return flight for a German tourist in Thailand is about 2.5 CO_2 equivalents, according to the carbon calculator of Carbon Footprint Ltd. A traveller can 'offset' this emission by buying 2.5 carbon credits. A number of companies have during the last few years arrived on the scene to sell such carbon credits. They take payments from travellers, pay a supplier of climate mitigation, and of course, charge a commission for themselves.

One of us participated in a large conservation conference organized by the International Union for Conservation of Nature (IUCN) in Barcelona in 2008. There were several thousand participants from many different countries. IUCN encouraged everyone to buy voluntary carbon credits for their travel emissions. These carbon credits were sold through the conference website. The result was proudly announced: This was a 'climate-neutral' conference.

The second type of carbon trade is one in which an industrialized company or country can buy carbon credits in a developing country and receive the right to register these in the climate account as reduction of carbon emissions for their own country. This is a mechanism that was established under the Clean Development Mechanism (CDM) of the Kyoto Protocol that was adopted in 1997 and entered into force in 2005.

Let us have a look at an early use of the carbon market. In 1993, the Dutch foundation 'Forest Absorbing Carbon Emissions' (FACE) started to sell carbon credits from forest plantations at the slopes of Mount Elgon in Uganda (Cavanagh and Benjaminsen 2014). According to the plan, the project should result in the capturing of 3.73 million tons CO_2 from 1994 to 2034, but already in 2003, the project fell apart. This was due to local resistance and conflicts because of forced evictions of smallholders in order to clear the area to make space for the planting of trees to capture carbon. At the same time, the project was presented on the webpage of FACE as a win-win project. This is an example of commodity fetishism that we introduced in ▶ Chap. 2.

In this case, carbon that was stored in the Ugandan forest, was sold to make up for emissions of greenhouse gases by people in the Netherlands. It was a Dutch energy company called N.V. Sep that established FACE in 1990. The original aim was that tree planting in Uganda should secure the company's rights to carbon emissions from a new coal power plant in the Netherlands. We can, in other words, conclude that this was case of climate injustice, where people in Mount Elgon lost their land and livelihoods because Dutch corporate interests wanted to mask their increases in CO_2 emissions through carbon trade.

Carbon is a virtual commodity that is detached from the context in which it is produced (Lohmann 2010). It is also an 'uncooperative' commodity in the sense that carbon is more difficult to measure and quantify than most other commodities (Bumpus 2011). In addition, forest projects are typically affected by the problems of 'leakage' and 'permanence'. The former refers to the possibility that deforestation may be moved to neighbouring forests, and the latter to the risk of stored carbon being released through fire, disease, pests and human encroachment (Cavanagh and Benjaminsen 2014).

CDM certification is not supposed to be approved if there are negative social impacts of a carbon capture project. There is, however, a question to what extent certification consultants are able to observe and detect serious problems. One of the tree plantations of the Norwegian company Green Resources in Uganda is CDM certified. However, researchers and journalists have criticized the project for leading to dispossession of local farmers and causing local conflicts (e.g. Lyons and Westoby 2014). The Swedish Energy Agency had a carbon credit agreement with Green Resources, which they terminated in 2020 due to the negative impacts that had been exposed.

The third type of carbon trade is sometimes called quota trading. As in the two other types, the trade concerns tons of CO_2 equivalents. The participating actors are companies that buy and sell carbon credits with each other.

The EU started with such a system in 2005 (*Emission Trading Scheme—ETS*), and it involves nearly half of the EU's climate emissions. Norway has participated in ETS since 2008, and about half of the carbon emissions in Norway in 2017 came from 140 companies in industrial sectors, which the government decided should take part in ETS. About 11,000 participating companies in ETS can buy and sell carbon credits to each other. A company that wants to emit more than the carbon credits it possesses, can do so if it purchases the necessary amount of carbon credits from other companies. On the other side, a company can carry out emission cuts and thereby sell their excess of carbon credits. The idea is that this will imply that the cheapest cuts will be made first, and that the money spent will result in the largest possible emission cuts (Norway's Ministry of Climate and Environment 2017).

ETS is based on 'cap-and-trade', which implies that a 'cap' in terms of amounts of carbon credits is gradually reduced. EU has decided to make a 2.2% reduction of carbon credits in ETS each year during 2021–2030. ETS is the largest system for quota trade. Other similar systems have been established in for instance California, Quebec, and in South Korea and Japan, while it is on its way in China.

Let us summarize the three forms of carbon trade and the various types of involved actors. First, voluntary markets imply that anybody on a voluntary basis can buy carbon credits from somebody who mitigates climate emissions, and in many cases, these mitigations are related to forests. The UN is in charge of the second type through CDM, and this is often trade between countries in the Global North as buyers, while sellers are enterprises in the Global South. One of the requirements for approval of this type of carbon trade is that the emission cuts or carbon capture in question are results of the carbon trade and would not otherwise have taken place. For countries in the Global North, purchasing such carbon credits can contribute to the fulfilment of the country's obligations according to international climate agreements. Quota trade constitutes the third type, and here the trade takes place between enterprises that all have large carbon emissions. These enterprises are required to be a part of the quota trade market, and a cap on the total number of carbon emissions is gradually lowered over the years, in order to gradually reduce the emissions of greenhouse gases. Countries like Norway use quota trade as a way to fulfil their international obligations for emission cuts. Apart from the buyers and sellers of carbon credits, carbon trade also involves a large apparatus of public and private actors who administer, regulate, investigate, control and assist the exchanges between buyers and sellers.

6.8 Discourses on Carbon Trade

Carbon trade consists of economic practices resulting from many decisions taken by governments and international organizations, and these decisions are based on a cluster of arguments that constitute a win-win discourse on climate trade. The main premise in this discourse is that carbon trade reduces emissions of greenhouse gases in a cost-effective way. When the trade involves emission cuts in the Global South, climate trade is also argued to create possibilities for revenues for poor countries, communities and people.

Behind the win-win discourse and creation of institutions for carbon trade, we find that governments of the US, the Netherlands, Canada and Norway have been particularly active. This is first of all a discourse elaborated by economists. It has also been quite visible in the international climate negotiations and in international organizations such as the World Bank and the UN programmes UNEP and UNDP.

In Norway, the win-win discourse on carbon trade has for a long time been hegemonic. Internationally, a radical discourse on carbon trade is produced by some of the same social and environmental organizations that are critical to REDD. A number of social scientists have also pointed at problematic aspects of carbon trade. The radical discourse entails arguments that carbon trade will not result in large enough cuts of greenhouse gases to avoid serious climate change. It is also seen as a vulnerable way to mitigate climate change in cases where forests are used as carbon sinks, because of measurement problems as well as problems with permanence and leakage. Instead, carbon trade is seen as 'green washing' of strategies based on continuous climate emissions by corporations, countries and consumers in the Global North (see e.g. Lohmann 2010; Gilbertson and Reyes 2009; Bond 2012; Bäckstrand and Lövbrand 2019).

More recently, however, a radical counter discourse has become influential in Norway, based on the main argument that larger carbon cuts should be made in Norway rather than paying for cuts in low-cost countries. Environmentalists have connected this argument to the question of new concessions for oil and gas prospecting and production outside the coast of Northern Norway in vulnerable areas of conservation value and with important fisheries.

The radical discourse on carbon trade questions whether this trade actually provides cost-effective climate mitigation as the proponents argue, since the conduct of carbon trade itself is expensive. In addition, a typical political ecology question in this context is to ask: Who is carbon trade cost-effective for?

6.9 Overarching Discourses on Climate Mitigation

As discussed in ▶ Chap. 3, in our article on environmental discourses from 2001, we identified three leading environmental discourses on four different issues: bioprospecting, forest conservation, desertification and climate change (Adger et al. 2001). Around 20 years later, we can still see that climate discourses at the international level contain basic elements of the win-win discourse, which emphasizes international action based on the principle of cost-effectiveness and with a large focus on carbon trade. The win-win discourses on both REDD and climate trade can be seen as examples of such discourses on specific themes. As the second climate discourse, Adger et al. (2001) identified what we here call a radical climate discourse that challenges the win-win-discourse, and we find such radical discourses also about REDD and climate trade.

The third climate discourse identified by Adger et al. (2001) expresses scepticism about the existence of serious and human-caused climate change, and provides arguments against the necessity of climate mitigation. This discourse belongs to the broader group of Prometheus discourses that oppose action to conserve the environment. This climate sceptic discourse has changed over the last few years owing to the strengthened evidence of global warming. There are currently few state leaders who still refuse the idea of serious anthropogenic climate change. Among the exceptions are the former US President Donald Trump, Poland's President Andrzej Duda and Brazil's President Jair Bolsonaro. All three defend their own countries' economic activities, such as the coal industry in the US and Poland, and expansion of agricultural activities that cause deforestation in the Amazonas.

6.10 The Case of an Oil-Fed Climate Change Discourse

Let us go back to the early history of climate policy formulation in Norway in order to understand how the foundation of the present policy was shaped. Hoven and Lindseth (2004) identified two discourses in Norwegian climate policy debates, and they called the one that came to dominate 'thinking globally'. This discourse is to a large extent in accordance with the international win-win discourse on climate change. They called the other discourse 'national action', which had a certain impact in the early 1990s, but was soon marginalized by 'thinking globally'. 'National action' is more or less in line with the radical climate discourse identified at the international level, as both emphasize the responsibility of the industrialized countries in the Global North to reduce their own climate emissions, and both discourses see climate change as a serious threat, which must be met with substantial mitigation measures.

Towards the end of the 1980s, it became increasingly clear that climate change constituted a serious global threat. In 1987, the Brundtland Commission recommended that the global consumption of fossil fuels be reduced by 50% over the next 30–40 years. Two years later, Norway with Brundtland as Prime Minister was one of the first countries to decide a target for reduction of climate emissions. This so-called stabilization target was to stop the growth of CO_2 emissions around year 2000 (Government of Norway 1988–89). However, this aim was soon replaced by a principle of *international cost-effectiveness*, which has been central to Norway's climate policies ever since, and it is based on the combination of two ideas. First, that climate emissions should be reduced where it is possible to get the largest cuts for the same amount of money. Second, that reduction of climate emissions is often likely to be cheaper in other countries, since Norway is one of the most expensive countries in the world. Emission cuts could take place in Europe, although low-income countries in the Global South might be seen as particularly relevant. Thus, climate mitigation must be carried out within international cooperation.

The principle of international cost-effectiveness was first used when the government decided to focus on the establishment of international markets for carbon trade. For a company in Norway, this implies that emissions requirements from the government can be met by buying carbon credits from a factory abroad that cuts its emissions instead of doing the same cuts itself. The Norwegian REDD programme from 2007 was based on the principle of international cost-effectiveness. Nevertheless, the programme was proposed by two Norwegian conservation organizations, and from the very start, it has been viewed as a programme not only to mitigate climate change, but also to conserve tropical forests. Thus, the programme must be seen as having been established through a coalition of the two interests of conserving forests and 'conserving' economic activities with large climate emissions.

Kasa (2016) shows how the Norwegian climate policy to a large extent was influenced by lobby activities from companies and labour unions in the petroleum sector and other large climate emitters. For instance, the former director of the company Hydro said in 1990:

> Since the most important environmental problems are global, it is now much more important to find solutions where all countries participate than to adopt increasingly stricter environmental requirements in little Norway, which anyway would have a small effect totally (Our translation from Norwegian) (Sæther 2017).

Nilsen (2001) presents a detailed record of individuals, research institutions and networks involved in the establishment of Norwegian climate policy in the early 1990s, with initiatives from the fossil fuel lobby with corporate and union leaders, and involving politicians and economists. Asdal (2014) shows how economists in the Ministry of Finance played a central role in changing Norway's climate policy to not threaten Norway's long-term interests of securing continued petroleum revenues. The institute for climate research, CICERO, was established not only to conduct research, but also to take a central role in helping establish an international network that promoted international cost-effectiveness and carbon trade as core elements of the international climate regime that developed around the negotiations and follow-up meetings of the UN's Climate Change Framework Convention (UNCCFC) (Nilsen 2001).

A successful lobby activity from Norway resulted in a wording in the UNFCCC before the 'Rio Earth Summit' in Rio de Janeiro in 1992, saying that the countries' commitments could be carried out in cooperation (Nilsen 2001). In 1997, the idea was brought into the Kyoto Protocol and, in 2015, into the Paris Agreement. Hence, the Norwegian fossil fuel lobby was successful in their aims of influencing the establishment of the Norwegian climate policy, and the international climate regime adopted similar ideas.

The financing of climate cuts in other countries has been the most important approach subscribed to by the Norwegian government since the early 1990s. This has made it possible for the government to present Norway internationally as well as to its own citizens as a leader in climate mitigation, while at the same time maintaining a high level of fossil fuel production.

6.11 Political Ecology from Hatchet to Seed Through Climate Justice

In ► Chap. 1, we introduced Robbins' (2020) distinction between political ecology as 'hatchet' and as 'seed', in other words, as critical and constructive analyses. In light of the climate crisis, we see a pressing need for 'seed' research to identify the best actions to curb global warming. Nevertheless, action-oriented research has to be based on solid knowledge of problems and causes. In order to sow, one first needs to use a hatchet to clear the ground.

In this chapter, we have discussed whether a country's climate change mitigation programme can be seen as temporal and spatial climate justice. When considering both the Norwegian context of the programme and evidence from the case study from Tanzania, we conclude that this may be seen as a case of *climate colonialism* (also called carbon colonialism by some) implying that land and natural resources in a foreign country are appropriated for climate mitigation through the use of financial power.

In line with scholarly work in general, critical and constructive analyses in political ecology will always be normative. However, political ecology is one of the fields where scholars aim to expose the values they bring to a discussion, instead of pretending their research is value-free. On climate change, we see a need for research that illuminates how various countries can play a positive role to counteract global warming in ways that provide temporal as well as spatial climate justice.

What would be the best choices for a country to cut climate emissions? This may vary between countries. In the case of Norway, we will mention three possible types of emission cuts that can be made in accordance with the two basic principles for climate justice.

The first option is that fossil fuel producing countries such as Norway initiate a close-down of this industry reducing emissions to zero as soon as possible. Such a policy would mean taking climate justice in time and space seriously.

Second, as we have shown in this chapter, forest-related climate interventions in the Global South cannot be seen as a just alternative. In order to ensure climate justice of such interventions, there need to be measures in place to avoid causing harm for forest adjacent communities. However, such measures would be expensive with costs such as real compensation, genuine participation by local communities in decision-making as well a serious and transparent system of knowledge production. But by taking such measures seriously, REDD could not anymore be seen as a cost-effective way of mitigating climate change. Hence, a 'cheap' and cost-effective REDD depends on there being a gap between a win-win discourse that pays lip-service to Social and Environmental Principles and Criteria (a UN framework often referred to) and practices that neglect these principles and criteria.

A third possible and even more ground-breaking way towards emissions cuts for a country like Norway would be to start elaborating alternatives to the neoliberal ways of organizing the economy. The most wealthy parts of the world are currently responsible for high and increasing levels of consumption, which are clearly not sustainable nor justifiable in a climate justice perspective. As an alternative, degrowth is increasingly receiving more attention in public debates as well as in the academic literature (see ► Chap. 2).

6

Conclusions

In this chapter, we have discussed how Norway as a fossil fuel producing nation has had a policy for the last 30 years of outsourcing its climate mitigation to so-called low-cost countries, because this is seen as more 'cost-effective'. Forest conservation in tropical countries to store carbon is an example of this, and when the REDD programme was launched in 2007, Norway soon became its biggest sponsor.

Furthermore, the chapter discusses the case of a Norwegian funded REDD project in Tanzania and how it had adverse effects on the livelihoods of people in the 21 villages around the two forest reserves that were highly restricted for local use as a consequence of the project. At the same time, the compensation for lost forest access that was provided mainly in the form of support to 'conservation agriculture' failed to generate sustained income for the farmers. The short time-period of the project (3 years of funding) combined with the harmful consequences for the villagers and a modest and failed compensation programme, demonstrate that neither the funder (Norway) nor the implementing NGO (African Wildlife Foundation) took the social impacts of this initiative seriously. At the same time, Norway has presented this particular project as a uniquely successful example of REDD implementation. Through the presentation of this and other forest-based climate interventions, Norway has been able to use its discursive power to support a win-win discourse about REDD as a highly successful climate mitigation programme.

In addition, outsourcing climate mitigation also depends on the existence of a functioning carbon market. The chapter discusses different forms of carbon trade and problematic aspects of this trade that may easily occur. Finally, we end up the chapter with discussing how a climate policy based on international cost-effectiveness could be replaced with policy guided by principles of climate justice.

Questions

1. What are the main sources contributing to climate change in your home country (or in another country you are interested in)?
2. How large is your country's total contribution of climate emissions at present? How have these emissions evolved so far, and what targets have been made for emission cuts in coming years?
3. Discuss climate justice of your country's total emissions.
4. Find some texts about your country's climate mitigation measures. Discuss similarities and differences in comparison with the discourses presented in this chapter.

References

Adger, W.N., T.A. Benjaminsen, K. Brown, and H. Svarstad. 2001. Advancing a political ecology of global environmental discourses. *Development and Change* 32 (4): 681–715.

Asdal, K. 2014. From climate issue to oil issue: Offices of public administration, versions of economics, and the ordinary technologies of politics. *Environment and Planning A* 46: 2110–2124.

Bäckstrand, K., and E. Lövbrand. 2019. The road to Paris: Contending climate governance discourses in the post-Copenhagen era. *Journal of Environmental Policy & Planning* 21 (5): 519–532.

Bond, P. 2012. *Politics of climate justice. Paralysis above, movement below*. Scottsville, South Africa: University of KwaZulu-Natal Press.

Bumpus, A.G. 2011. The matter of carbon: Understanding the materiality of tCO2e in carbon offsets. *Antipode* 43 (3): 612–638.

Büsher, B. 2014. Selling success: Constructing value in conservation and development. *World Development* 57: 79–90.

Caney, S. 2014. Two kinds of climate justice: Avoiding harm and sharing burdens. *Journal of Political Philosophy* 22 (2): 125–149.

Cavanagh, C., and T.A. Benjaminsen. 2014. Virtual nature, violent accumulation: The 'spectacular failure' of carbon offsetting at a Ugandan National Park. *Geoforum* 56: 55–65.

———. 2015. Guerrilla agriculture? A biopolitical guide to illicit cultivation within an IUCN Category II protected area. *Journal of Peasant Studies* 42 (3–4): 725–745.

Chapin, M. 2004. A challenge to conservationists. *World Watch Magazine*, November/December, 17–31.

Eckstein, H. 1975. Case studies and theory in political science. In *Handbook of political science. Political science: Scope and theory*, ed. F.I. Greenstein and N.W. Polsby, vol. 7, 94–137. Reading: Addison-Wesley.

Flyvbjerg, B. 2006. Five misunderstandings about case-study research. *Qualitative Inquiry* 12 (2): 219–245.

Gilbertson, T., and O. Reyes. 2009. Carbon trading: How it works and why it fails. *Critical currents. Dag Hammarskjöld Foundation Occasional Paper Series*, pp. 7–29. http://www.thecornerhouse.org.uk/sites/thecornerhouse.org.uk/files/OscarTamCarbonTrade.pdf.

Government of Norway. 1988–89. *Stortingsmelding (White Paper) No. 46*.

Government of Norway. 2018. Norway's REDD+ disbursement. https://www.regjeringen.no/en/topics/climate-and-environment/climate/climate-and-forest-initiative/kos-innsikt/how-are-the-funds-being-spent/id734170/. Visited 28 Apr 2020.

Hoven, E., and G. Lindseth. 2004. Discourses in Norwegian climate policy: National action or thinking globally? *Political Studies* 52: 63–81.

IPCC. 2018. *Special report: Global warming of 1.5 °C*. United Nations: International Panel on Climate Change. https://www.ipcc.ch/sr15/.

Jansen, E.G. 2009. *Does aid work? Reflections on a natural resources programme in Tanzania*. Christian Michelsen's Institute, U4 Issue 2009: 2, www.U4.no.

———. 2014. Don't rock the boat: Norway's difficulties in dealing with corruption in development aid. In *Corruption, grabbing and development: Real world challenges*, ed. T. Søreide and A. Williams. London: Edward Edgar.

Kasa, S. 2016. Klimautfordringer i et norsk og skandinavisk perspektiv. In *Det norske samfunn*, ed. Ivar Frønes and Lise Kjølsrød, vol. 1, 7th ed. Oslo: Gyldendal.

Lohmann, L. 2010. Uncertainty markets and carbon markets: Variations on Polanyian themes. *New Political Economy* 15 (2): 225–254.

Lund, J., E. Friis, M. Sungusia, B. Mabele, and A. Scheba. 2017. Promising change, delivering continuity: REDD+ as conservation fad. *World Development* 89: 124–139.

Lyons, K., and P. Westoby. 2014. Carbon colonialism and the new land grab: Plantation forestry in Uganda and its livelihood impacts. *Journal of Rural Studies* 36: 13–21.

Mung'ong'o, C.G., C. Lyamchai, I. Kulaya, P. Mushi, G. Sayula, and M. Semlowe. 2011. *Establishing baseline conditions in the Kolo Hills Forests and adjacent communities*, A Final Draft Socio-Economic Baseline Survey Report. Dar es Salaam: African Wildlife Foundation.

Munsterhjelm, N. 2019. *Mapping vegetation change in REDD+ pilot projects in Tanzania using optical satellite imagery time series*. BSc thesis, Faculty of Science and Technology, Norwegian University of Life Sciences.

Nilsen, Y. 2001. *En felles platform? Norsk oljeindustri og klimadebatten i Norge fram til 1998*. PhD thesis, University of Oslo.

Nixon, R. 2011. *Slow violence and the environmentalism of the poor*. Cambridge: Harvard University Press.

Norway's Ministry of Climate and Environment. 2017. Klimakvoter. http://www.regjeringen.no/no/dep/kld/id668/.

Popper, K. 1959. *The logic of scientific discovery*. New York: Basic Books.

Putnina, S. 2019. *Conservation through intensification: Adoption of agricultural technologies introduced by a REDD project in Kondoa, Tanzania*. Master's thesis, Faculty of Landscape and Society, Norwegian University of Life Sciences.

Ritchie, H., and M. Roser. 2017/2019. CO_2 and greenhouse gas emissions. *Our World in Data*. https://ourworldindata.org/co2-and-other-greenhouse-gas-emissions#cumulative-co2-emissions. Visited 30 June 2020.

6

Robbins, P. 2020. *Political ecology. A critical introduction*. 3rd ed. Oxford: Wiley Blackwell.

Royal Norwegian Embassy in Dar es Salaam. 2012. *Norway & Tanzania – Partners in development*. Dar es Salaam: Royal Norwegian Embassy.

———. 2014. Payment for preservation of forests. www.norway.go.tz.

Sæther, A.K. 2017. *De beste intensjoner: Oljelandet i klimakampen*. Oslo: Cappelen Damm.

Svarstad, H. 2021. A critical climate education: Studying climate justice in time and space. *International Studies in Sociology of Education* 30(1). https://doi.org/10.1080/09620214.2020.1855463.

Svarstad, H., and T.A. Benjaminsen. 2017. Nothing succeeds like success narratives: A case of conservation and development in the time of REDD. *Journal of Eastern African Studies* 11 (3): 482–505.

Thunberg, G. 2019. *No one is too small to make a difference*. London: Penguin Books.

United Nations. 2015. *World fertility patterns 2015. Data booklet*. New York: United Nations.

UN-REDD Programme. 2012. *Social and environmental principles and criteria*. UN-REDD Programme Eighth Policy Board, March 25–26, 2012. Asunción, Paraguay. UNREDD/PB8/2012/V/1.

Woods Hole Research Center. 2013. Measuring the role of deforestation in global warming. http://www.ucsusa.org/global_warming/solutions/stop-deforestation/deforestation-global-warming-carbon-emissions.html#.V6CMXWVPJVo.

World Commission on Environment and Development (WCED). 1987. *Our common future*. Oxford: Oxford University Press.

Pastoralists and the State

Contents

T. A. Benjaminsen, H. Svarstad, *Political Ecology*,
https://doi.org/10.1007/978-3-030-56036-2_7

Trailer

In 2016, the Sámi reindeer owner Jovsset Ánte Sara from Finnmark in the far north of Norway took the Norwegian state to court. As a young and newly established livestock-keeper, Sara aims to build up a sizeable herd that can sustain him economically. But the Norwegian state's policy of bringing down overall reindeer numbers tries to force him to reduce his herd to 75 animals from the 116 he previously had. He claims this is a decision that in practice means the end of his activities as a pastoralist, since it is impossible to make a living from 75 reindeer. He therefore went to court in order not to lose his livelihood.

In March 2016, Sara won the first round in the Primary Court, but the state appealed to the Secondary Court in Tromsø where Sara also won. The state, however, appealed again and in December 2017, the Supreme Court in Oslo ruled against Sara. While he appealed to the United Nations Human Rights Committee in Geneva, the state immediately started to implement its destocking policy through administrating heavy fines for not following the court's ruling combined with plans for forced slaughter of Sara's reindeer.

This case illustrates how Sámi reindeer-keepers in Norway are squeezed by a policy to reduce both the numbers of reindeer and their owners. It is also an example that demonstrates the marginalization that Sámi reindeer-owners have been subject to in Norwegian society for a couple of centuries. But such marginalization is not unique to reindeer-keepers. It is found in most countries where there are pastoralists. We will in this chapter take a closer look at this relationship between pastoralists and the state (◘ Fig. 7.1).

◘ **Fig. 7.1** Jovsset Ánte Sara's sister, Máret Ánne Sara, is an artist. This is how she depicts destocking forced upon Sámi reindeer herders by Norwegian authorities. The process, which curiously is framed as part of a policy of 'self-determination', also creates competition among herders. (Source: Máret Ánne Sara)

Definition

Pastoralists are livestock-keepers who move with their animals, over long or shorter distances, to find pasture and water. Often some members of the household live in one place, while others herd the livestock part of the year. In other cases, but this is rarer today, the whole household may move in a nomadic adaptation.

Overview

Globally, there are somewhere between 100 and 200 million pastoralists. They tend to have a poor reputation among modern state authorities. Pastoralism is in fact often seen as an unproductive vestige of the past and as a stage in development somewhere between hunter-gatherers and agriculture. As a consequence of this view, governments frequently attempt to turn pastoralists into sedentary farmers. The vast majority of countries that have pastoralists within their borders follow such a policy. Modern governments have a tendency to view nomadic use of resources as messy and illegible (Scott 1998), because nomadic migrations do not fit into a modern form of governmental organization. Pastoralists are difficult to tax and control, and it is also difficult to adapt services such as schools and health care to their mobility.

This is, however, not only a Western perception, or one reserved for capitalist economies. Also in China, and in the countries that make up the former Soviet Union, pastoralists tend to be seen as primitive and unmanageable. Permanent settlement and making livestock keeping more efficient and sustainable is therefore considered the solution to the problems that mobility represents for governments.

This view on pastoralism represents what we may call a *modernization discourse*. State authorities tend to feel a need to actively control and govern pastures and pastoralists through quantifiable indicators. The scientific basis for this policy approach is found in succession theory within range science that is usually referred back to the American ecologist Frederic Clements (1874–1945). He presented his theory in two volumes (Clements 1916, 1920), which in brief says that without 'disturbance', usually thought of from humans or livestock, ecosystems will naturally go through various stages of succession before reaching climax, which is the system's natural condition. Grazing will disturb this process. It is therefore important that the grazing pressure stays within a calculated carrying capacity, which is the threshold indicating how much outtake the system can take before it is tipped out of balance. Clements developed this theory based on field studies on the prairie in Nebraska, which is in a temperate climate. He claimed, however, that this was a general theory that was applicable to all climatic zones, and the theory was quickly adopted by public administration in the USA as the scientific basis for the management of pastures all over the country (Sayre 2017).

Even if it soon turned out that the succession theory was less suitable to describe the dynamics of the drier and warmer grazing areas in for instance Arizona and New Mexico, the model still dominated American range management until the 1990s. This can be explained by the fact that it responded to a bureaucratic need of governing through quantitatively defined variables such as carrying capacity (i.e.

how many animals a defined grazing area can sustain). The success of succession theory as a management tool is therefore more a result of a bureaucratic and political need for quantifiable parameters rather than being an expression of universally accepted scientific conclusions (Sayre 2017).

It was only in the 1970s and 1980s that succession theory started to be challenged by alternative ideas about ecosystems, especially in drylands, and how they are unstable and not in equilibrium, contrary to what Clements' theory said. It is therefore meaningless to define a carrying capacity in such systems. This non-equilibrium theory was developed based on ecological research in Australia, Israel and Africa. In addition, anthropologists and economists, who studied pastoral societies in Africa, played an important role in formulating the theory (Sayre 2017). This theory development culminated with a conference in Woburn in England in 1990, which resulted in a seminal book (Behnke et al. 1993) that has become a standard reference on non-equilibrium theory used on pastoral areas.

Today, most researchers on pastoral systems agree that succession theory (or equilibrium theory) gives a good description of processes in temperate grazing systems, but that non-equilibrium theory better captures the dynamics in areas with more seasonal and annual variations, in particular Tropical drylands, but also for instance cold mountainous areas and the Arctic.

Pasture management in the USA has since this scientific paradigm shift in the 1990s adopted these new ideas (Sayre 2017). This management is in addition inspired by ideas about adaptive management (Walters and Holling 1990) based on a stepwise and flexible approach in contrast to forms of top-down management associated with how succession theory tends to be implemented. However, in most other parts of the world, succession theory and the modernization discourse dominate policy formulation.

In this chapter, we will present a few such examples—from African drylands as well as from reindeer husbandry in the Arctic. In particular, we will discuss two main arguments about the environmental impact and economic contribution of pastoral systems, which are the key assumptions behind the modernization discourse about pastoralism.

First, pastoral policies are generally guided by the idea that pastoralism tends to result in overgrazing if left out of the state's control. Second, pastoral production is often claimed to be inefficient, unproductive and contribute little to national economies. We will take a closer look at these arguments and the critique against them. In this way, this chapter also discusses two classical themes in political ecology—degradation and marginalization.

7.1 Debates About Overgrazing

The paradigm shift in range science from equilibrium to non-equilibrium ecology that started in the 1990s has also inspired a new *flexibility discourse* about pastoralism. In an African context, it emphasizes that African savannas have a high *resilience*. In other words, the natural environment has an inherent ability to recover after a shock such as a drought. In addition, arid regions constitute unstable systems that are dominated by one factor in particular—the annual rainfall, which is usually subject to considerable fluctuations from year to year.

People who depend on natural resources in such arid regions have therefore learned that the only certainty is uncertainty (Berge 2000). In such an environment, it is necessary to be flexibly adaptive to survive. Moreover, uncertainty is not unique to Africa, but is found in all environments where pastoralists live. These are usually locations with marginal resources, whether in drylands, highlands or Arctic regions, which means that the renewable resources that are used are often unstable and distributed thinly over large areas (with some exceptions of key resources such as wetlands in drylands). For centuries, pastoralists have developed various ways of adapting to a changing resource base, which range from a purely nomadic life (which is rare nowadays) to what is called 'transhumance'. Transhumance means a fixed base from which herders and livestock move for a shorter or longer part of the year. This migration takes place in order to find pasture and water in a different ecological zone, while the rest of the household stays in the base.

Definition

Sandford (1983) wrote the first influential publication to criticize what he called the *Mainstream View on pastoral development*. This was also one of the first theoretical contributions to the new flexibility discourse, while Ellis and Swift (1988) presented the first substantial empirical study to undermine the mainstream view based on field data from 9 years of ecological research in northern Kenya. According to Ellis and Swift (1988), the mainstream view (or modernization discourse) has been based on three assumptions: that African pastoral ecosystems are potentially stable (that a carrying capacity can be identified), that these potentially stable ecosystems are frequently destabilized by improper use on the part of pastoralists, and that alterations of system structure (reducing livestock numbers, changing land tenure patterns etc.) are needed to return these systems to equilibrium and a more productive state. Hence, the idea is that if ecosystems are left undisturbed, they will evolve naturally towards a climax stage. Grazing disturbs these systems and, if exceeding the carrying capacity, will lead to degradation. In order to prevent such degradation, it is argued that those responsible for management (usually a governmental unit), must determine the number of animals the system can support. In order to be able to preserve the ecosystem at an acceptable stable level, the number must not exceed this carrying capacity.

Ellis and Swift (1988) demonstrated, however, that it does not make sense to fix a carrying capacity in a dry environment dominated by unstable rainfall. Moreover, they concluded that it is the climate - more than the density of livestock - that determines biomass and its composition and distribution in drylands.

Since the early 1980s, research undertaken by the French ecologist Pierre Hiernaux in northern Mali has also played an important role as an empirical basis for the flexibility discourse (Hiernaux 1993; Hiernaux et al. 2009, 2016; Mougins et al. 2009; Dardel et al. 2014; Benjaminsen and Hiernaux 2019). Hiernaux's data material represents an unusually long and detailed data series in African ecological drylands research.

The northern Sahel is an area with especially high resilience. Due to the domination of annual grass species in the Sahel and their strong seasonal dynamics, the risk of overgrazing is limited to a short period in time. This critical period is in the growing season before the seeds fall to the ground and can be as short as 2 weeks. In fact, the two basic properties of Sahelian pastoral ecosystems, instability and resilience, support the continued practice of transhumance in the southern Sahel and of nomadism in the northern Sahel. Both these methods of utilizing resources are based on mobility and maximal dispersion of livestock during the growing season. For example, Tuareg nomads in Northern Mali move their camps approximately every third day in the rainy season. This is because they seek *em-n-amsud* (the tip of the grass), which is said to be preferred by the cattle (Benjaminsen 1997).

Definition

Theoretically, one may distinguish between two types of carrying capacity in an equilibrium system—ecological and economic (Behnke and Scoones 1993). This also leads to a theoretical divide between two types of *overgrazing*. Ecological overgrazing is defined as a threshold, which the grazing intensity should not exceed to avoid negative effects on the regeneration of pastures, while overgrazing in economic sense implies that the pastures cannot satisfy the nutritional need of the livestock. This need is again dependent on the production objective—whether it is meat production for a market or milk production for own consumption (Homewood and Rodgers 1987; Benjaminsen 1997) or whether the aim is maximization of carcass weight or herd productivity per area unit (Benjaminsen et al. 2006; Marin et al. 2020).

► Example

A commonly shared idea about the impact of grazing in African drylands has been that high grazing pressure around wells leads to overgrazing. To test this hypothesis, Hanan et al. (1991) analysed biomass production around boreholes in the Ferlo region in Senegal. During 30–35 years, the areas around these wells had been exposed to heavy grazing. If this kind of pressure on the vegetation leads to degraded pastures, it should be detectable in this area. Biomass production up to 25 km from 20 wells in the region was studied by ground data measurements and the analysis of meteorological satellite images at the end of the rainy seasons in 1987 and 1988. The conclusion was that there was no consistent relationship between biomass production and distance to the wells, and therefore no evidence of long-term degradation. In other words, the grass cover near the wells grew just as fast as the grass further away. However, Hanan et al. (1991) did not study changes in species composition as a result of grazing pressure. Evidently, grazing may favour some species at the expense of others.

More recently, Rasmussen et al. (2018) also investigated the impact of grazing around boreholes in the Ferlo region in Senegal using satellite images. These authors found 'inverse grazing gradients' in terms of decreasing biomass production with distance from the wells. This can be explained by the effect of livestock fertilization of soils at higher livestock densities closer to the wells. ◀

These results are similar to those obtained by a team of geographers from the University of Lund studying an area in Sudan. They concluded: 'There was no trend in the creation or possible growth of desert patches around 103 examined villages and water holes over the period 1961–1983' (Helldén 1991, p. 379). Other studies using satellite images have demonstrated the resilience of the vegetation cover in the Sahel. Tucker et al. (1991) and Tucker and Nicholson (1999) showed for example that what they defined as the desert boundary was in fact dynamic and fluctuated along a north-south axis from year to year in response to the annual rainfall.

The flexibility discourse is presently generally accepted by scholars with regard to African savannas, although the modernization discourse still dominates among many decision-makers, in Africa and internationally. Among scientists there is, however, a debate about to what extent other regions and ecosystems in the world with pastoralism also behave as unstable systems that are mainly controlled by climatic factors.

Livestock Keeping and Ecological Change in South Africa

In South Africa, there has been a debate in particular about ecological change and management of the Karoo region in the Northern Cape province (Illius and O'Connor 1999; Todd and Hoffman 1999; Sullivan and Rohde 2002; Riginos and Hoffman 2003; Benjaminsen et al. 2006; Allsopp et al. 2007). This is a dry area, but with relatively stable winter rainfall. The area is dominated by succulent plants, and will generally have a very different ecological dynamic than for example the grass-covered savannas in the Sahel. The Karoo is highly valued by conservationists because of its many endemic species. Of the 4849 recorded plant species in the succulent Karoo, 1940 are endemic and subject to a rapid loss of habitat, according to Myers et al. (2000).

The debate about the ecological change on farmland in South Africa is indeed also a political debate involving differing views about what the landscape should look like, what is appropriate land use, and which priorities are most important. South Africa is one of the countries in the world with the most unequal land distribution. The white minority, which today constitutes about 10% of the population, still owns about 70% of agricultural land, despite land reform. A large part of the black majority lives in communal areas (formerly called Homelands and Bantustans, and Coloured Reserves in the Western and Northern Cape Provinces). These areas constitute approximately 13% of agricultural land. The communal areas are remnants from the colonial and apartheid era and are still owned by the state. In these reserves, people were kept in poverty through land scarcity in order to serve as a reserve of cheap labour for the mining industry and the white farmers (see also ► Chap. 4).

Photo 7.1 Fence-line contrast. (Photo: Anke Hoffmann)

7

In today's debate on the management of communal areas in South Africa, the idea of the 'tragedy of the commons' (Hardin 1968) holds a strong position. Communal areas are particularly among politicians and government officials associated with overstocking and the small-scale farmers' use of these areas is seen as outdated. They are told to learn from white large-scale 'commercial' farmers and adopt their methods. The densities of people and livestock are naturally far greater (at least 10 times as high) in communal areas compared to the private large-scale farms, and this naturally has consequences for vegetation and landscape. But the question is whether an intensively used landscape is necessarily overgrazed and degraded.

Landscape or aerial photographs of so-called fence-line contrasts are often used as evidence of overgrazing and degradation on the communal side. These are photos of areas divided by fences between communal areas and private farms. Photo 7.1 depicts a spectacular fence-line contrast, which, according to Hoffmann and Zeller (2005), shows an 'overgrazed communal farming area' on the left and a 'moderately grazed' farming area on the right.

By comparing aerial photographs of such fence-lines before and after the fences were put up in Namaqualand, Benjaminsen et al. (2006) questioned the presentation of evidence in fence-line photos. On an aerial photo taken in 1960 in Namaqualand, it is impossible to tell the difference between a communal area and the neighbouring private farm (Photo 7.2). The density of livestock and impact on vegetation was about the same for private farms and communal areas. Fences were then put up at the beginning of the 1960s by the apartheid government. In addition to receiving subsidies for the development of infrastructure (wells, dams, fencing), white farmers received subsidies to reduce the number of livestock on their farms. The idea was to reduce grazing pressure, increase carcass weight and produce meat of better quality for the market.

In the next photo, taken in 1997, one can clearly identify the fence-line between the two properties as a result of decreased stocking density on the private farm (Photo 7.3). Hence, there has been a recovery of vegetation on the farm, while in the communal area, the vegetation cover appears to be approximately the same as in

Photo 7.2 Aerial photograph taken in 1960 of the border between Leliefontein communal area and a private farm in Namaqualand. Livestock densities were about the same on the two properties. (Source: Department of Land Affairs)

Photo 7.3 Aerial photograph taken in 1997 of the same area as in Photo 7.2. One sees a clear fence-line contrast between the two properties. Vegetation has increased on the private farm due to subsidies from the government to reduce livestock numbers to produce better quality meat for the market. (Source: Department of Land Affairs)

the first photograph. Instead of documenting overgrazing in the communal area, one may therefore conclude that there is a certain inherent resilience in this predominantly succulent vegetation system. In addition, the photographs indicate that a continued loss of vegetation in the communal area has not occurred, although the area gives the impression of being worn down compared to the neighbouring farm.

Comparison of fluctuations in livestock numbers over time supports the idea of a resilient system. Table 7.1 presents the changes in animal numbers over time in

Table 7.1 Small Stock Units (SSU) in Leliefontein communal area by year. Note the reduction in SSU due to multi-year droughts between 1903–1907 and 1998–2000. One SSU equals a sheep or a goat

Year	Smallstock units
1824	24,000
1854	25,230
1875	24,400
1890	25,600
1909	14,700
1947	34,890
1953	40,000
1997	36,479
2002	23,712

Source: Benjaminsen et al. (2006)

the Leliefontein communal area in the Northern Cape. The figures show that grazing pressure has been around the current level for about 180 years. If there had been a continuous deterioration of pastures—ecological overgrazing—it would not have been possible to maintain this number of livestock over time. In that case, one would have expected a decrease in animal numbers. Table 7.2 from the Concordia communal area also demonstrates that the number of livestock is correlated with the annual rainfall. This supports the idea in the flexibility discourse that the pastoral grazing system to a large extent is driven by climatic factors.

It is also interesting to note that there is still a large species diversity in this area, in spite of several hundred years of intensive livestock-keeping. The persistent high density of livestock in the communal areas has produced an old cultural landscape with a different biological diversity than what is found for areas where use is less intensive. This is a cultural landscape that is not valued in southern Africa today. Agricultural authorities tend to compare the use of the land in communal areas with how the white-owned commercial farms are run, while conservationists argue for the reinstatement of a pristine landscape and seem to prioritize the protection of biodiversity at the cost of poverty alleviation (Benjaminsen et al. 2008).

The idea that African small-scale farming degrades the landscape is used by different influential actors in the discussion of land reform in southern Africa. First, it is used as an argument against redistribution of land in general (Hongslo and Benjaminsen 2002). Second, this idea serves to justify prioritizing so-called emergent black commercial farmers, as opposed to more widespread redistribution (Benjaminsen et al. 2006). Third, it is used as an argument for placing land under the

Table 7.2 Small stock units in Concordia and rainfall (mm per year) in Springbok, 1909, 1938, 1971–1988, 2002–2003

Year	SSU	Residual (absolute value)	Year-on-year Change SSU	Rainfall 1-year lag	Rainfall 2-year mean lagged
1909	18740	1323		211.2	170.00
1938	19327	308		161.4	177.05
1971	14768	3062		136.6	201.85
1972	4002	13792	10766	199.7	168.15
1973	10211	7547	6209	164.5	182.10
1974	15497	2224	5286	134.4	149.45
1975	31539	13854	16042	218.1	176.25
1976	22016	4367	9523	192.3	205.20
1977	23362	5749	1346	335.4	263.85
1978	38942	21365	15580	239.5	287.45
1979	17567	26	21375	106.0	172.75
1980	10965	6540	6602	115.6	110.80
1981	11925	5544	960	189.8	152.70
1982	20982	3549	9057	178.1	183.95
1983	23436	6039	2454	199.8	188.95
1984	20951	3590	2485	276.4	238.10
1985	13475	3850	7476	109.4	192.90
1986	12129	5160	1346	199.1	154.25
1987	11233	6020	896	208.6	203.85
1988	16672	545	5439	171.9	190.25
2002	17917	1204		151.4	194.25
2003	12235	4442	5682	192.7	208.55

Source: Benjaminsen et al. (2006)

auspices of the national park agency (SANParks) in competition with redistribution to landless people (see ► Chap. 4).

In other words, this debate is clearly marked by the values and interests of the most influential actors. Agricultural authorities use white commercial farms as a model and ideal for the development of farming in communal areas. Therefore, they require that the stocking density must be reduced and adapted to a fixed carrying capacity, which is economically defined and based on the aim of maximizing carcass weight meaning to maximize meat production per animal. In order to have 4–5-months-old lambs ready for slaughter, it is necessary to have abundant pastures available. Therefore, the stocking density cannot be high.

In the communal areas, however, which in Namaqualand are generally located in the same ecological zones as private farms, many small-scale farmers share grazing land with a high density of livestock on the land, but with relatively few animals per farmer. This means that it is unrealistic for these farmers to aim only at meat production, which requires large properties in this dry environment. Therefore, communal farmers do not only have the production of meat as an objective for livestock-keeping. Saving for future unforeseen expenses is probably the most important objective. Very few of the small-scale farmers have livestock as their only source of income. Those who are older receive a small pension from the government. Others are miners, farm workers on commercial farms or have other paid work in urban areas.

Claims about overgrazing can of course also be empirically correct from a particular ecological perspective. Intensive grazing in the communal areas has obviously impacted on the species composition, and some species have decreased in prevalence locally, while others have been favoured by grazing. For example, annual species increase on areas that are grazed heavily. Some see this type of change as a clear sign of overgrazing. However, we have in the presentation of this case study from South Africa tried to show how discussions about overgrazing should not be confined to technicalities within ecology or agronomy, but must be seen in a broader perspective that includes issues of economics, livelihoods, rights, justice and values.

Sámi Reindeer Pastoralism in Northern Norway

Discussions about Sámi reindeer herding and overstocking in Finnmark in Northern Norway have many similarities with the debate in South Africa. Here, we also find both the modernization discourse and the flexibility discourse. Reinert (2006) says that the former is rooted in the idea of Fordist mass production, while the latter is based on the idea of Arctic pastoralism adapting to its environment. The modernization discourse on reindeer herding in Norway is produced by politicians, government officials, actors proposing new mining or windmill projects, environmentalists and the media. It says quite simply that through a logic described by the tragedy of the commons model there are too many reindeer in Finnmark, which results in overstocking and environmental degradation.

Reindeer husbandry is the primary livelihood for over 20 ethnic groups in the circum-polar Arctic and Sub-Arctic area. In Norway, Finnmark in the far north is the largest reindeer-herding region, with 150,000–200,000 reindeer and about 2150 reindeer owners of Sámi ethnicity (Benjaminsen et al. 2016). Across most of Norway, reindeer herding is reserved for people of Sámi ethnicity, and access to this industry is regulated through a system of licenses to restrict the number of pastoralists.

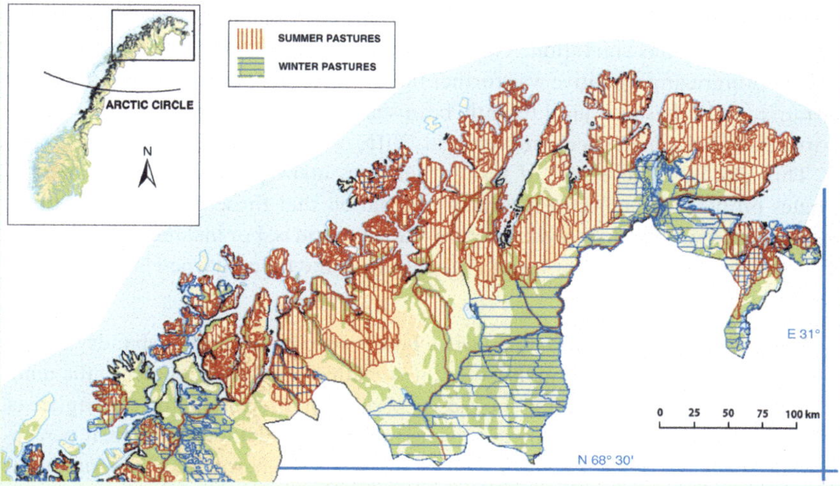

Fig. 7.2 Summer and winter pastures in Finnmark, Northern Norway. (Source: Benjaminsen et al. 2015)

Reindeer pastoralism is based on a pattern of seasonal migrations. In Finnmark, these migrations take place between the mountainous winter pastures of the inland, which are dominated by lichen, and the green summer pastures on the coast (Fig. 7.2). Annual migrations from winter to summer pastures can extend as far as 200–300 km in each direction.

Since reindeer husbandry is practiced on 40% of the land area in Norway, there are a number of land-uses that compete with this activity, such as mining, windmill parks, municipal and state-led development of infrastructure, recreation activities and conservation of carnivores. These competing land-uses have strong advocates, for example, in the Parliament, who see reindeer herding as an obstacle, which needs to be reduced in order to continue to develop Norwegian society. On the other hand, reindeer husbandry has hardly ever been represented in the Parliament.

Nevertheless, the notion that Sámi reindeer herders keep an excess of animals is not new. Since the end of the nineteenth century, the state has periodically aimed at reducing reindeer populations. Initially, this was principally justified by the need to minimize land-use conflicts between pastoralists and sedentary farmers. However, during the last few decades, the size of the reindeer population has been restated as being primarily an environmental problem, a matter of overstocking, and the ecological impact of excessive reindeer populations. With the increased focus on environmental issues in the 1970s and 1980s, terms, such as 'overgrazing', moved to the fore in Scandinavian research on reindeer management (Pape and Löffler 2012). Along with this increased attention, the matter of excessive reindeer was also progressively configured as a technical problem, to be solved through managerial interventions (Reinert 2012; Johnsen and Benjaminsen 2017).

In the late 1990s, there were some winters with difficult snow and ice conditions. This meant that reindeer could not access pastures through deep snow and ice, and the consequence was starvation. This received much attention in Norwegian media, and the overgrazing narrative was further fuelled. Several researchers supported this narrative, and it has also until recently been shared by all the political parties in the Norwegian Parliament (Benjaminsen et al. 2016).

This lack of achievement of results frustrates many politicians. Parliamentary debates on the issue reflect a recurring resignation that those involved in reindeer husbandry do not understand how serious the situation is. For instance, in the debate of November 30, 2000, Minister of Agriculture at the time, Bjarne Håkon Hansen, expressed this frustration:

> What concerns me is the fact that when I meet with the reindeer husbandry groups and discuss the reality, and insist that if something does not happen to the number of reindeer, there will be the risk of permanent destruction of grazing areas, then my perception of reality is on a completely different level than their perception of reality. It is *not* a grazing problem, insists the Norwegian Association of Reindeer Herders, it is a predator problem, it is a land-use issue, and there have been some special natural conditions that have resulted in the pastures becoming "locked". But it is not a grazing problem - there are adequate pasture resources. It is clear that this assembly and I can take responsibility, and we can use the law to provide us with the tools to first fix a maximum number of reindeer per district, and then go further to put in place legislation that gives us the opportunity to fix a maximum number of reindeer per production unit. I think this is the only way to ensure that one will obtain a decrease in the number of reindeer. But we also know that the industry, represented by the Norwegian Association of Reindeer Herders and the Sámi Parliament, are in disagreement with this. Then, we face such a serious problem that to some extent, I actually feel incapable of acting.

In June 2000, the Government asked the Norwegian Parliament to establish a maximum number of reindeer for each district. In December the same year, the Ministry of Agriculture followed this up by asking the Reindeer Management Board, which is an office under the Ministry, to implement this measure. Based on a commissioned study, the highest number of reindeer for West Finnmark was set at 64,300 animals. This is the biggest reindeer herding area and also seen by the government as the most troublesome. The proposed number was adopted by the Reindeer Management Board in 2002. In 2005, however, the number of animals reached around 91,000 in West Finnmark. This was the topic of the parliamentary debate on June 14, 2005 where several members of parliament across the political spectrum pointed out that the voluntary reduction of reindeer numbers clearly did not work, and argued for forced reductions.

Hence, there is a political alliance to reduce the number of reindeer in Finnmark. Reindeer husbandry is not treated as a business, but as an activity where outdated ideology and attitudes continue to live (Reinert 2006). While record production in any other agricultural activity would be subject to celebration, high production of reindeer is viewed as a problem (Reinert 2006).

Biologists at the University of Tromsø and the Norwegian Institute of Nature Research (NINA) in Tromsø have for a couple of decades been providing the scientific justification for the policy of reducing reindeer numbers based in particular on the claim that there are too many reindeer on the winter pastures. Their research is an example of how studies based in the natural sciences may not only be normative, but manifestly political, while the scientists themselves continue to pretend that their work is objective and apolitical. Hence, this is an example of apolitical ecology (see ► Chaps. 1 and 2), which in reality is just as political as political ecology, if not more so.

Among the explicit recommendations from this group of biologists is that participation and co-management involving the Sámi reindeer owners does not lead to better management. They criticize the 'myth' that 'co-management, voluntary agreements, and economic incentives (in reindeer husbandry), so-called "carrot-incentives", are more efficient than the good old whip' (Økland 2010). In this way, these biologists situate themselves within a long Norwegian colonial history in relation to the Sámi people, and in particular in relation to reindeer pastoralism where the use of the whip (or the stick, and not the carrot) has been the rule rather than the exception.

This is research, which is adapted to, and which reproduces the dominant discourse in Norwegian society about Sámi reindeer pastoralism, which is further implemented through the state's reindeer governance. This top-down governance has since about 2010 gradually developed into a policy of forced reduction of reindeer numbers and threats of exceedingly high fines for the reindeer owners who do not comply. In this way, these biologists also return the favour to the Ministry of Food and Agriculture for generous research funding over many years. Using this research as justification, the ministry then responds by swinging the whip over a small minority of Sámi reindeer pastoralists. So, clearly reindeer governance and research on reindeer management in Norway is in urgent need of decolonization.

The reindeer-keepers themselves have, however, a completely different understanding of Arctic pastoral landscape dynamics. They argue that it is mainly climatic factors (in addition to encroachments from other land-uses) that determine the survival and condition of their reindeer, which corresponds to a large extent to the ideas found in non-equilibrium thinking and the flexibility discourse. As seen for instance in the quote above by the agricultural minister, government officials and politicians tend, however, to treat this as a problem of misinformation among the pastoralists further reinforcing colonial perspectives and top-down management including lack of recognition of traditional knowledge (Benjaminsen et al. 2015).

This experience-based knowledge that reindeer herders have acquired over many generations has in particular a focus on fluctuations and unpredictability. For instance, mild weather with subsequent periods of frost can form ice layers. The reindeer can break through one such layer of ice to get down to the lichen below, but several layers or large amounts of snow may be too much. Then the pastures will be 'locked' under ice and snow and the reindeer will starve—regardless of how much pasture is available below the layer of snow. This situation occurred in the winter and spring of 2020 when unusual amounts of hard packed snow made pastures inaccessible to reindeer, which had to be fed with feed concentrate. So, while the reindeer suffered from this situation, the lichen and other vegetation under the snow remained largely untouched.

This means that precipitation and temperature throughout the winter are important factors governing access to pasture, and not only the number of animals (Marin 2003; Benjaminsen et al. 2015; Marin et al. 2020). The anthropologists Bjørklund (1990) and Paine (2004) have made important contributions to this flexibility discourse. Paine was sharp in his criticism of the Norwegian government's focus on the maximum number of reindeer, which he said is written in stone, and does not take into account the instability of tundra ecology.

In Sámi, there are many concepts that refer to the well-being of reindeer and the condition of pastures throughout the year. The concept of *guohtun*, for example is often used to mean 'the availability of pastures through snow' (Eira et al. 2010; Eira 2012). This makes it problematic to translate *guohtun* simply as 'pasture' as the Reindeer Herding Administration tends to do. So, when the herders say that *guohtun* is bad this winter (meaning that the access to the pastures is difficult because of the amounts of snow and ice), this has been translated in government documents to mean 'bad pastures' or 'overgrazed pastures'. But in bad winters with starving reindeer (which is quite rare because herders would often give additional feed in these cases such as hay or pellets as in 2020), the pastures do not suffer, because they are hardly grazed.

The indigenous knowledge of the reindeer herders gives in fact a completely different image of reindeer herding than presentations in many scientific publications (although there are scientific publications supporting this knowledge as well), by politicians, government officials and the media. In practice, however, the knowledge of the reindeer pastoralists remains largely invisible and marginalized in public debates in Norway.

Scientifically, the reindeer overstocking narrative is to a large extent based on time-series of satellite images of the lichen layers in Finnmark. These time-series show a reduction in the area covered by lichen in the 1980s and 1990s. In their analysis of reindeer pastures, Johansen and Karlsen (2005) used satellite images from 1973, 1980, 1987, 1996 and 2000. They found that the lichen had covered one third of the area in 1973 and 1980, but—following a steady decrease—only 6% in 2000. During the same period, other vegetation types increased. Reduction in the quantity of lichen coincided with an increase in the reindeer population from the 1970s to around 1990. Although reindeer numbers decreased again in the 1990s (Fig. 7.4), the consensus among many scientists is that increased reindeer numbers are the principal cause of the observed reduction in lichen (e.g. Riseth et al. 2004; Johansen and Karlsen 2005; Hausner et al. 2011).

A report by Johansen et al. (2014) reconfirmed these views strengthening public and political opinion in Norway about the need to destock reindeer populations, especially in Finnmark. The main conclusion in the report, widely publicized in the media, was that the lichen had covered 19% of the winter pastures in 1987, but only 4% in 2013. The ecological crisis on the winter pastures was confirmed, and there was urgent need to adopt swift political action. This has further led to the implementation of a strict policy of quotas for each reindeer herding district and individual reindeer owners based on calculated carrying capacities. Exceeding these quotas leads to heavy fines and forced slaughter down to the required number of animals.

However, this study by Johansen et al. (2014) is flawed in several ways. First, the study involved a number of questionable normative assessments. For instance, one of the mapped land-use categories is labelled 'worn lichen pastures'. These are lichen pastures above the tree line, and no dramatic changes in this category were detected over time. Yet, surprisingly, this category was not included in the estimates of lichen changes by Johansen et al. (2014). There is also the question of what 'worn' implies. What are the conditions that these pastures are being compared to? Or, what is an acceptable amount of lichen and who decides what this amount should be? This is after all an old pastoral landscape in which reindeer pastoralism has been practiced for hundreds of years. Therefore, one should naturally expect a pastoral impact on the vegetation, meaning that this is not an untouched wilderness.

In addition, and even more problematically, the loss of lichen cover only refers to a category called 'lichen ground'. These are open areas below the tree line dominated by lichen. It is quite logical that this land-use category has decreased as there is encroachment of bushes and trees in such areas in most Norwegian mountains. Tømmervik et al. (2009) found, for instance, a doubling of birch in this area between 1957 and 2006. This is an ongoing process of landscape change all over Norway, which is usually attributed to climate change leading to warmer and wetter summers (Aas 1969; Bryn 2008) combined with reduced grazing by livestock in some areas. Despite this, in discussions of landscape change in Sámi pastoral areas, this conclusion seems to be forgotten, and instead overstocking of reindeer is blamed.

Based on the satellite data, simplified colour maps are made to illustrate the threat of overstocking. ◘ Figure 7.3 is such a map, which is reproduced from the most recent report of the office of the Auditor General (Riksrevisjonen 2012: 51). In this figure, the complexities of the pastoral landscape are reduced to a rough colour map, with a large section in red labelled 'overgrazed' and a small, ever shrinking part in white labelled 'intact'. The connotative purpose of using red is obviously to indicate danger (see ► Chap. 2 on the connotative meaning of images). However, one wonders what the baseline is of the notion of 'intact' being invoked here: an 'original' landscape from the 1970s, which would have remained exactly the same without reindeer pastoralism?

This map is powerful as a political tool and is often used by ministers or government officials eager to demonstrate that there is an overabundance of reindeer in comparison to the grazing resources available. ◘ Figure 7.3 efficiently communicates the perceived danger of overstocking.

In his seminal book 'How to lie with maps', Monmonier (1991) points out that maps are often used in political propaganda:

> » A good propagandist knows how to shape opinion by manipulating maps… The propagandist molds the map's message by emphasizing supporting features, suppressing contradictory information, and choosing provocative, dramatic symbols. People trust maps, and intriguing maps attract the eye as well as connote authority. Naïve citizens willingly accept as truth maps based on a biased and sometimes fraudulent selection of facts (p. 87).

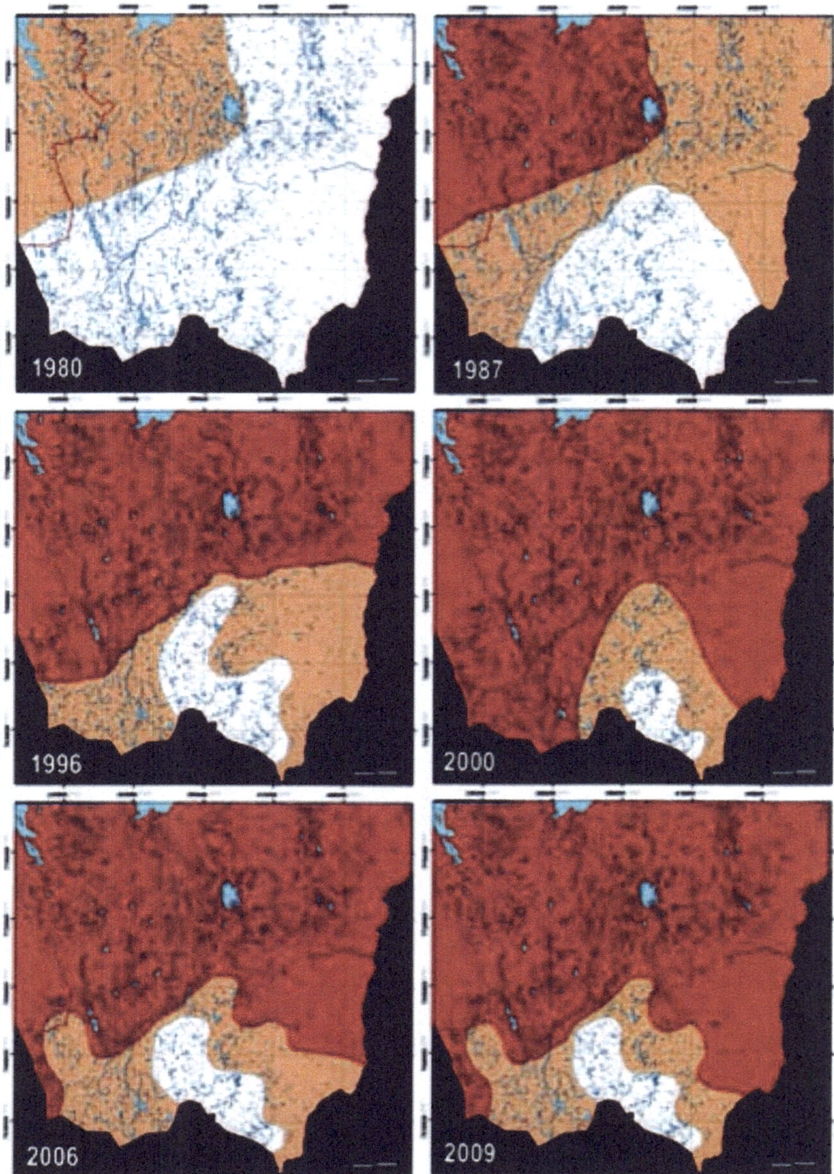

Fig. 7.3 Overgrazing on the winter pastures in Finnmark according to the Auditor General. The map is based on time series from Johansen and Karlsen (2005). Red = overgrazed, orange = moderately grazed; white = intact. (Source: Riksrevisjonen 2012)

7

Furthermore, Monmonier (1991) adds that 'most propaganda maps are little more than cartoons' (p. 113). In particular, one should be suspicious when contrasting hues are used to demonstrate differences in intensity on choropleth maps. ◘ Figure 7.3 is a lucid example of a map that has been simplified to the level of a cartoon and used as scientific evidence in the Norwegian state's political propaganda to reduce Sámi reindeer pastoralism to a minimum (Benjaminsen 2020).

Nevertheless, the ecological monitoring carried out by the Norwegian Institute of Nature Research (NINA) since 1998 is the most reliable empirical research available on the impact of reindeer populations on the winter pastures in Finnmark. Results from the first comparative analysis (1998–2005) show that lichen cover had a marked increase from 1998 to 2005, from an average of 18.3% to 27.6% (Gaare et al. 2006). Additionally, the thickness of the lichen layers increased from an average of 22 mm to 28 mm during the same period. Gaare et al. (2006: 5) conclude that 'today climate factors appear to be determinant for the continued development of the balance between reindeer and the pastures'. It is interesting to note that this conclusion and the positive increase in pastures in Finnmark occurred at the same time as the reindeer population of West Finnmark increased between 2000/2001 and 2005 (◘ Fig. 7.4).

The second analysis based on the monitoring of pastures expands the time series until 2010 concluding that, during the period 2005–2010, there had been an overall decline in lichen coverage from 27.1% to 24.5% as well as a change in the average

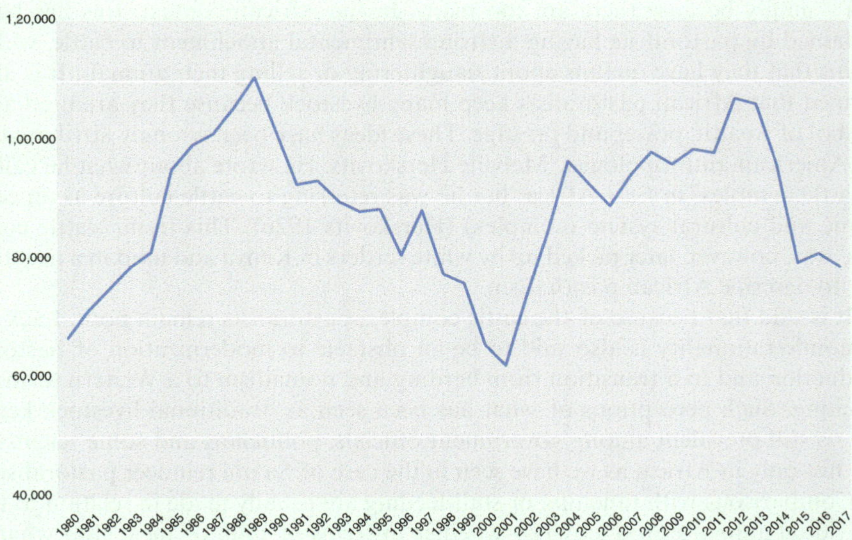

◘ **Fig. 7.4** Reindeer numbers in Western Finnmark, 1980–2017. (Source: Marin et al. 2020)

thickness of the lichen layer from 29 mm to 23 mm (Tømmervik et al. 2011). This occurred during a period when the reindeer population had continued to increase. Despite this decline, the lichen layer increased in both extension and thickness for the whole period investigated (1998–2010).

Nevertheless, it is not surprising that the amount of lichen is reduced with more reindeer, although the relationship between reindeer and lichen is not linear, and climate variation also clearly plays a role. Hence, grazing inevitably affects vegetation with regard to both species composition and distribution, but despite the high reindeer numbers in 2010, there was no ecological crisis on the tundra. This conclusion stands in contrast to the public and policy-based image given of reindeer husbandry in Norway leading to overgrazing, which is depicted in ◘ Fig. 7.3. The lack of scientific evidence for the overstocking narrative testified by the conclusions from this ecological monitoring and the problems with the analyses of satellite imagery, combined with the massive media and policy obsession with this narrative, demonstrate that the narrative has taken the form of a myth in Norwegian society (Benjaminsen et al. 2016; Benjaminsen 2020).

7.2 Debates About Economic Irrationality Among Pastoralists

It has long been the prevailing view that pastoralists keep unnecessarily high numbers of animals, which in addition to causing overgrazing are also in poor condition—simply because there are too many. In the African context, this has been explained by pastoralists having a strong sentimental attachment to cattle, which means that they have qualms about slaughtering or selling their animals. It is also claimed that African pastoralists keep many livestock because they are used as a symbol of wealth, power and prestige. These ideas have been wrongly attributed to the American anthropologist, Melville Herskovits. He wrote about what he called a 'cattle complex' in East Africa, but he was referring to cattle culture as an economic and cultural system (complex) (Herskovits 1926). This term, 'cattle complex' was, however, later picked up by white settlers in Kenya and used in a negative way to describe African pastoralism.

It is said that because of the cattle complex, pastoralists remain poor. Lack of economic rationality is also said to be an obstacle to modernization of pastoral production and to a transition from herding and nomadism to a Western form of ranching. Such perceptions of what has been seen as 'traditional livestock keeping', is still prevalent among government officials, politicians and some scientists, and not only in Africa, as we have seen in the case of Saami reindeer pastoralism.

Comparisons with ranching or stall-feeding are usually made in relation to the individual animal's productivity—whether in terms of milk or meat. But what is not taken into account then is that pastoralists maximize production per area instead of per individual animal as opposed to stall-feeding and ranching. This is simply a necessity, because they live in marginal and variable environments, where migrations and the tracking of environmental variability are necessary. This approach is also important in order to maximize the number of people who may subsist in such marginal environments. The alternative to this system is privatization of large areas for ranching for a few, such as in South Africa. This would lead to the marginalization of large groups of people who may be excluded and end up in urban areas as slum-dwellers.

For pastoralists, maximizing the number of animals is also a strategy to buffer the fluctuations in access to pasture and food. While farmers store grain from one harvest to the next, it is too cumbersome for herders to transport grain or fodder. Therefore, they save by having larger herds than they strictly need in the short term in order to sell or exchange animals for grain from farmers during periods of scarcity (Benjaminsen and Berge 2004). This is usually the time of the year when livestock prices are low. But for pastoralists, it is more important that the animals can move themselves, than receiving a maximum price at an unfavourable time. In African drylands, the rainy season is an unfavourable time of year to sell animals. This is primarily because this time is when the animals produce the milk that is the most important source of carbohydrates and proteins, and because during the rainy season, livestock often reside far from the markets. In addition, investments in livestock in normal and good years bring a large return, even though such investments also involve the risk that one may have large losses during droughts.

Table 7.3 below compares the productivity found on ranches in the United States and Australia and among African pastoralists under comparable climatic conditions. It indicates that land productivity may be greater in areas with pastoral use than in Western ranching. The table suggests that African pastoral systems can be up to ten times as productive per unit area as ranches in the USA and Australia in similar arid regions.

The contribution of pastoral livestock to national economies in African countries is also often underestimated (Hesse and MacGregor 2006). This may be because there is no political will to acknowledge this contribution, but also because such informal economic activity represents a measurement problem.

Table 7.3 Livestock productivity in different regions and systems under comparable climatic conditions

Region, system	Protein produced (kg/hectare/year)
USA, ranching	0.3–0.5
Australia, ranching	0.4
Sahel, West Africa, herding	0.6–3.2
Kenya, Masai group ranches, herding	3.4
Karamojong, Uganda, herding	3.6
Borana, Ethiopia, herding	2.1

Source: Grandin (1987)

Within Sámi reindeer pastoralism in Norway, there is a situation similar to what is described in ◘ Table 7.3. Since pressure from other competing land-uses has historically been more intense in the southern reindeer herding areas, the result has been a dramatic reduction in the number of pastoralists and animals, and a confinement of herding to a limited area with short migrations. Hence, in these southern areas, around Røros south of Trondheim, the system is approaching the ranching model. There is still herding, or 'guarding' rather, because the reindeer are followed by the pastoralists at a distance, often using binoculars, and the pastoralists move in the landscape with the use of ATVs or snowmobiles. This distance between reindeer and pastoralists mean that the reindeer remain wild, or semi-wild.

Because of a milder climate, less movements with shorter migrations, lower reindeer densities, and more abundant feed, individual reindeer in southern Norway are fatter than the sturdy and smaller reindeer in the far north. It is these southern areas that are presented by the Ministry of Food and Agriculture as the model for other reindeer pastoralists in Norway, while Western Finnmark is seen as the main problem area where the reindeer owners to a lesser degree follow the advice from the government (Benjaminsen et al. 2016; Johnsen and Benjaminsen 2017). This means that fewer calves are slaughtered than what the government recommends and that the share of bucks in the herd is higher than recommended by state advisors.

According to the state's reindeer policy, it is only necessary to have a few bucks in the herd for reproduction purposes. Having many bucks would mean less pasture available for the valuable female reindeer that produce calves. Based on experiences from sheep production where winter feed is the limiting factor, the state recommends slaughtering of calves in the autumn that are born in spring the same year. This means that the herds are reduced to a minimum before winter, and only females are kept in addition to a few bucks for reproduction. In this way, a surplus of lamb or calves is harvested every year.

The purpose with this model is therefore to optimize meat production, and to make production as efficient as possible. In addition, the purpose is to reduce pressure on the winter pastures that are believed to be degraded or at least be a limiting factor.

However, reindeer owners point out that the bucks have many tasks in the herd in addition to reproduction. They contribute to better use of pastures by grazing areas close to human habitation and infrastructure that are not used by the more easily scared females and calves. The muscle power of the bucks, and especially the castrates, is also useful during winter to dig through layers of snow and ice to access the pastures. In this way, bucks and castrates help to make pastures accessible for the rest of the herd. To have a relatively high percentage of bucks can therefore represent a form of insurance against difficult winter conditions when the pastures are 'locked'. A herd with few bucks may therefore be more vulnerable to climatic fluctuations (Benjaminsen et al. 2016).

It is interesting to note that productivity per area unit is, on average, approximately the same in West Finnmark (22.4 kg/km^2) and in the Røros area (21.8 kg/km^2) for the period 1980–2016 (◘ Fig. 7.5). This productivity is consider-

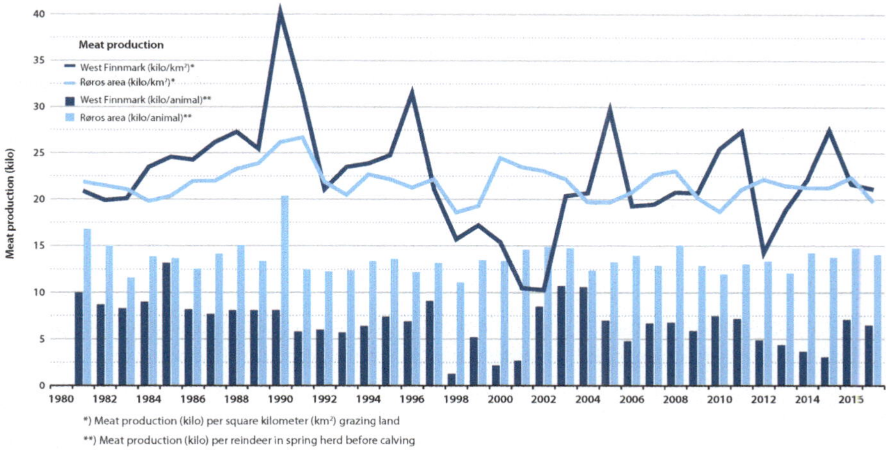

Fig. 7.5 Pastoral productivity in the Sámi reindeer herding areas Røros and West Finnmark measured per animal (bars) and per km^2 (graph). (Source: Norwegian Agriculture Agency and Johnsen and Benjaminsen (2017))

ably higher than in the other reindeer herding areas where the average is 15.1 kg/km^2 even if the Arctic conditions in Finnmark are more difficult with more climatic instability than what is found further south. This instability is also reflected in the fluctuations in production found in West Finnmark.

The pastoral contribution to national economies in various African countries is also often underestimated (Hesse and MacGregor 2006). This may be because there is not political will to acknowledge this contribution, but also because it is difficult to measure it. Nevertheless, pastoral production is estimated to represent 10% of Kenya's Gross Domestic Product (GDP), which is approximately the same as the agricultural contribution to GDP (Hesse and MacGregor 2006).

In Finnmark, reindeer pastoralism has changed from being a profitable activity in the 1970s to becoming a subsidized economic activity (Reinert 2006). Reindeer meat is after all a quality product produced outdoors with free-ranging animals consuming natural feed, and with a limited production. Nevertheless, profitability in reindeer-keeping fell dramatically from the 1970s to around 2010. For instance, the price of one kilogram of meat fell from about 13 US dollars in 1976 to about 6 US dollars in 1990, before increasing to 8 US dollars in 2011 (measured in the equivalent value of Norwegian kroner in 2011) (Reinert 2016). The main reason for this negative development from the 1970s was that reindeer-keeping was squeezed into a planning model where prices were fixed by the state and not in the relationship between supply and demand. This model was based on ideas about what is efficient production inspired by Norwegian farmers' production of cattle (Reinert 2016).

As long as the reindeer-keepers were able to govern their own business until the late 1970s, that is as long as they had a fair amount of food sovereignty (see ▶ Sect. 2.4 in ▶ Chap. 2), they managed to run a profitable business. But when the state took over with its planning model influenced by stationary forms of livestock-keeping, without understanding the importance of natural fluctuations, prices were allowed to drop when the production of meat increased, but when the production went down again, prices stayed low. The market had become a monopsony with one big dominant buyer who had an economic interest in keeping the payments to the producers down (Reinert 2016). Due to widespread critique, some cracks in this monopsony has emerged since 2012 combined with increased efforts to market reindeer meat, which have led to increased prices to producers again.

Conclusions

Pastoralists sustain a living by using land in areas that are too marginal or unstable climatically and ecologically to obtain sustainable agricultural production. These areas are typically highly productive during parts of the year and resource scarce in the rest of the year. Examples of such regions are African savannas and the Arctic tundra. Because the grazing resources of these regions are unstable, unpredictable, fluctuating and distributed thinly over large areas, the use of these resources must be flexible and based on movement of livestock—often over long distances. Such mobility fits poorly together with the will of modern governments to control land governance and populations.

Similarly, mobile resource management stands in contrast to the principles of sedentary agricultural production and will often be underestimated. In this way, the modernization discourse neglects the traditional knowledge of pastoralists. This discourse continues to dominate the understanding of most decision-makers on pastoralism. Consequently, the modernization discourse contributes to the political and economic marginalization of pastoralists, who operate in ecologically marginal regions.

The modernization discourse on pastoralism builds on ecological succession theory that was developed in the USA early in the twentieth century. There was, however, a paradigm shift in international research on pastoralism in the 1980s and 1990s. While range management in the USA seems now to have changed to follow the knowledge in the non-equilibrium model, in most African states as well as in the governance of reindeer pastoralism in Norway, the old paradigm based on equilibrium theory and often associated with a top-down approach still dominates planning and decision-making. It is this flawed governance that the young reindeer pastoralist Jovsset Ánte Sara does not accept, and that he has taken to court to fight.

This chapter also illustrates two central themes in political ecology—a deconstruction of the idea of environmental degradation caused by small-scale and subsistence-oriented users, and how these already marginalized users are subject to further marginalization through persistent attempts to modernize their production. Both are classic themes in political ecology. The first arises from critiques of neo-Malthusianism and the second comes from the influence of Marxism on political ecology.

Questions

1. Sum-up the main points of the Succession Theory vs. the Non-Equilibrium Theory.
2. Is an intensively used landscape necessarily overgrazed and degraded? Take the example of livestock-keeping in South Africa.
3. Describe the relationship between reindeer pastoralists and the Norwegian state.
4. What would a livestock policy that promotes environmental sustainability and social justice look like?

References

Aas, B. 1969. Climatically raised birch lines in southeastern Norway 1918-1968. *Norwegian Journal of Geography* 23 (3): 119–130.

Allsopp, N., C. Laurent, M.C. Debeaudoin, and M.I. Samuels. 2007. Environmental perceptions and practices of livestock keepers on the Namaqualand Commons challenge conventional rangeland management. *Journal of Arid Environments* 70 (4): 740–754.

Behnke, R., and I. Scoones. 1993. Rethinking range ecology: Implications for rangeland management in Africa. In *Range ecology at disequilibrium: New models of natural variability and pastoral adaptation in African savannas*, ed. R. Behnke, I. Scoones, and C. Kerven, 1–30. London: Overseas Development Institute & International Institute for Environment and Development.

Behnke, R.H., I. Scoones, and C. Kerven, eds. 1993. *Range ecology at disequilibrium. New models of natural variability and pastoral adaptation in African savannas*. London: Overseas Development Institute & International Institute for Environment and Development.

Benjaminsen, T.A. 1997. Natural resource management, paradigm shifts and the decentralization reform in Mali. *Human Ecology* 25 (1): 121–143.

———. 2020. Depicting decline: Images and myths in environmental discourse analysis. *Landscape Research*: 1–15. https://www.tandfonline.com/doi/full/10.1080/01426397.2020.1737663#.

Benjaminsen, T.A., and G. Berge. 2004. Myths of Timbuktu – From African El Dorado to desertification. *International Journal of Political Economy* 34 (1): 31–59.

Benjaminsen, T.A., and P. Hiernaux. 2019. From desiccation to global climate change: A history of the desertification narrative in the West African Sahel, 1900-2018. *Global Environment* 12: 206–236.

Benjaminsen, T.A., R.F. Rohde, E. Sjaastad, P. Wisborg, and T. Lebert. 2006. Land reform, range ecology, and carrying capacities in Namaqualand, South Africa. *Annals of the Association of American Geographers* 96 (3): 524–540.

Benjaminsen, T.A., T. Kepe, and S. Bråthen. 2008. Between global interests and local needs: Conservation and land reform in Namaqualand, South Africa. *Africa* 78 (2): 221–244.

Benjaminsen, T.A., H. Reinert, E. Sjaastad, and M.N. Sara. 2015. Misreading the Arctic landscape: A political ecology of reindeer, carrying capacities, and overstocking in Finnmark, Norway. *Norsk Geografisk Tidsskrift* 69 (4): 219–229.

Benjaminsen, T.A., I.M.G. Eira, and M.N. Sara, eds. 2016. *Samisk reindrift, norske myter*. Bergen: Fagbokforlaget.

Berge, G. 2000. *In defence of pastoralism: Form and flux among Tuaregs in northern Mali*. PhD dissertation, Department of Social Anthropology, University of Oslo.

Bjørklund, I. 1990. Saami reindeer pastoralism as an indigenous resource. *Development and Change* 21 (1): 75–86.

Bryn, A. 2008. Recent forest limit changes in south-east Norway: Effects of climate change or regrowth after abandoned utilisation? *Norwegian Journal of Geography* 62: 251–270.

Clements, F. 1916. *Plant succession. An analysis of the development of vegetation*. Washington, DC: Carnegie Institute.

———. 1920. *Plant indicators: The relation of plant communities to process and practice*. Washington, DC: Carnegie Institute.

Dardel, C., L. Kergoat, E. Mougin, P. Hiernaux, M. Grippa, and C.J. Tucker. 2014. Re-greening Sahel: 30 years of remote sensing data and field observations (Mali, Niger). *Remote Sensing of Environment* 140: 350–364.

Eira, I.M.G. 2012. *Muohttaga jávohis giella: Sámi árbevirolaš máhttu muohttaga birra dálkkádatrievdamis*. The Silent Language of Snow: Sámi Traditional Knowledge of Snow in Times of Climate Change. PhD dissertation, Department of Linguistics, University of Tromsø.

Eira, I.M.G., O.H. Magga, and N.I. Eira. 2010. Muohtatearpmaid sisdoallu ja geavahus. *Sámi dieđalaš áigečála* 2/2011, pp. 3–24. (Sámi snow terminology – Meaning and usage.)

Ellis, J.E., and D.M. Swift. 1988. Stability of African pastoral ecosystems: Alternate paradigms and implications for development. *Journal of Range Management* 41 (6): 450–459.

Gaare, E., H. Tømmervik, J.W. Bjerke, and D. Thannheiser. 2006. *Overvåking av vinterbeiter i Vest-Finnmark og Karasjok: Ny beskrivelse av fastrutene*. Norsk Institutt for Naturforskning, NINA Rapport 204.

Grandin, B.E. 1987. *Pastoral culture and range management: Recent lessons from Maasailand. International Livestock Center for Africa, Addis Ababa*. ILCA Bulletin 28, September.

Hanan, N.P., Y. Prevost, A. Diouf, and O. Diallo. 1991. Assessment of desertification around deep wells in the Sahel using satellite imagery. *Journal of Applied Ecology* 28: 173–186.

Hardin, G. 1968. The tragedy of the commons. *Science* 162: 1243–1248.

Hausner, V.N., P. Fauchald, T. Tveraa, E. Pedersen, J.L. Jernsletten, B. Ulvevadet, R.A. Ims, N.G. Yoccoz, and K.A. Bråthen. 2011. The ghost of development past: The impact of economic security policies on Saami pastoral ecosystems. *Ecology and Society* 16 (3): 4.

Helldén, U. 1991. Desertification – Time for an assessment? *Ambio* 20 (8): 372–383.

Herskovits, M.J. 1926. The cattle complex in East Africa. *American Anthropologist* 28: 230–272, 361–388, 494–528, 633–664.

Hesse, C., and J. MacGregor. 2006. *Pastoralism: Drylands' invisible asset? Developing a framework for assessing the value of pastoralism in East Africa*. IIED Issue Paper 142.

Hiernaux, P. 1993. *The crisis of Sahelian pastoralism: Ecological or economic?* Addis Ababa: International Livestock Centre for Africa.

Hiernaux, P., E. Mougin, L. Diarra, N. Soumaguel, F. Lavenu, Y. Tracol, and M. Diawara. 2009. Sahelian rangeland response to changes in rainfall over two decades in the Gourma region, Mali. *Journal of Hydrology* 375 (1–2): 114–127.

Hiernaux, P., C. Dardel, L. Kergoat, and E. Mougin. 2016. Desertification, adaptation and resilience in the Sahel: Lessons from long term monitoring of agro-ecosystems. In *The end of desertification? Disputing environmental change in the drylands*, ed. R.H. Behnke and M. Mortimore. Heidelberg: Springer.

Hoffmann, A., and U. Zeller. 2005. Influence of variations in land use intensity on species diversity and abundance of small mammals in the Nama Karoo, Namibia. *Belgian Journal of Zoology* 135: 91–96.

Homewood, K., and W.A. Rodgers. 1987. Pastoralism, conservation and the overgrazing controversy. In *Conservation in Africa. People, policies and practice*, ed. D. Anderson and R. Grove. Cambridge: Cambridge University Press.

Hongslo, E., and T.A. Benjaminsen. 2002. Turning landscapes into 'nothing': A narrative of land reform in Namibia. *Forum for Development Studies* 29 (2): 321–347.

Illius, A.W., and T.G. O'Connor. 1999. On the relevance of nonequilibrium concepts to semi-arid grazing systems. *Ecological Applications* 9: 798–813.

Johansen, B., and S.R. Karlsen. 2005. Monitoring vegetation changes on Finnmarksvidda, Northern Norway, using Landsat MSS and Landsat TM/ETM+ satellite images. *Phytocoenologia* 35: 969–984.

Johansen, B., H. Tømmervik, J.W. Bjerke, and T.R. Karlsen. 2014. *Finnmarksvidda – kartlegging og overvåking av reinbeiter*. Status 2013. NORUT Rapport 07/2014.

Johnsen, K.I., and T.A. Benjaminsen. 2017. The art of governing and everyday resistance: 'Rationalization' of Sámi reindeer husbandry in Norway since the 1970s. *Acta Borealia* 34 (1): 1–25.

Marin, A. 2003. *Protean reality: Discourses for managing the winter reindeer ranges in Finnmark, northern Norway*. MSc dissertation, Department of International Environment and Development Studies, Norwegian University of Life Sciences.

Marin, A., E. Sjaastad, T.A. Benjaminsen, M.N. Sara, and J. Borgenvik. 2020. Productivity beyond density: A critique of management models for reindeer pastoralism in Norway. *Pastoralism* 10 (1). https://pastoralismjournal.springeropen.com/articles/10.1186/s13570-020-00164-3#citeas.

Monmonier, M. 1991. *How to lie with maps*. Chicago: Chicago University Press.

Mougin, E., P. Hiernaux, L. Kergoat, M. Grippa, P. De Rosnay, F. Timouk, V. Le Dantec, V. Demarez, F. Lavenu, M. Arjounin, T. Lebel, N. Soumaguel, E. Ceschia, B. Mougenot, F. Baup, F. Frappart, P.L. Frison, J. Gardelle, C. Gruhier, L. Jarlan, S. Mangiarotti, B. Sanou, Y. Tracol, F. Guichard, V. Trichon, L. Diarra, A. Soumaré, M. Koité, F. Dembélé, C. Lloyd, N.P. Hanan, C. Damesin, C. Delon, D. Serca, C. Galy-Lacaux, J. Seghieri, S. Becerra, H. Dia, F. Gangneron, and P. Mazzega. 2009. The AMMA-CATCH Gourma observatory site in Mali: Relating climatic variations to changes in vegetation, surface hydrology, fluxes and natural resources. *Journal of Hydrology* 375 (1–2): 34–51.

Myers, N., R.A. Mittelmeier, C.G. Mittelmeier, G.A.B. Da Fonseca, and J. Kent. 2000. Biodiversity hotspots for conservation priorities. *Nature* 403 (6772): 853–858.

Økland, T.E. 2010. For mange rein. http://forskning.no/naturressursforvaltning-miljoovervakning-dyreverden/2010/02/mange-rein.

Paine, R. 2004. Saami reindeer pastoralism: Quo vadis? *Ethnos* 69 (1): 23–42.

Pape, R., and J. Löffler. 2012. Climate change, land use conflicts, predation and ecological degradation as challenges for reindeer husbandry in northern Europe: What do we really know after half a century of research? *Ambio* 41: 421–434.

Rasmussen, K., M. Brandt, X. Tong, P. Hiernaux, A.A. Diouf, M.H. Assouma, C.J. Tucker, and R. Fensholt. 2018. Does grazing cause land degradation? Evidence from the sandy Ferlo in Northern Senegal. *Land Degradation and Development* 2018: 1–11.

Reinert, E. 2006. The economics of reindeer herding: Saami entrepreneurship between cyclical sustainability and the powers of state and oligopolies. *British Food Journal* 108 (7): 522–540.

Reinert, H. 2012. The disposable surplus – Notes on waste, reindeer and biopolitics. *Laboratorium* 4: 67–83.

Reinert, E. 2016. Årsaker til at planøkonomien er blitt varig ødeleggende for verdiskapningen i samisk reindrift. In *Samisk reindrift, norske myter*, ed. T.A. Benjaminsen, I.M.G. Eira, and M.N. Sara, 145–174. Bergen: Fagbokforlaget.

Riginos, C., and M.T. Hoffman. 2003. Changes in population biology of two succulent shrubs along a grazing gradient. *Journal of Applied Ecology* 40: 615–625.

Riksrevisjonen. 2012. Riksrevisjonens undersøkelse av bærekraftig reindrift i Finnmark. *Dokument* 3: 14. (2011–2012).

Riseth, J.Å., B. Johansen, and A. Vatn. 2004. Aspects of a two-pasture-herbivore model. *Rangifer* 15: 65–81.

Sandford, S. 1983. *Management of pastoral development in the third world*. New York: John Wiley & Sons.

Sayre, N.F. 2017. *The politics of scale. A history of rangeland science*. Chicago: University of Chicago Press.

Scott, J.C. 1998. *Seeing like a state: How certain schemes to improve the human condition have failed*. New Haven: Yale University Press.

Sullivan, S., and R.F. Rohde. 2002. On non-equilibrium in arid and semi-arid grazing systems. *Journal of Biogeography* 29: 1–26.

Todd, S., and M.T. Hoffman. 1999. A fence-line contrast reveals effects of heavy livestock grazing on plant diversity and community composition in Namaqualand, South Africa. *Plant Ecology* 142: 169–178.

Tømmervik, H., B. Johansen, J.Å. Riseth, S.R. Karlsen, B. Solberg, and K.A. Høgda. 2009. Above ground biomass changes in the mountain birch forest and mountain heath of Finnmarksvidda, northern Norway, in the period 1957–2006. *Forest Ecology and Management* 257: 244–257.

Tømmervik, H., B. Johansen, S.R. Karlsen, and P.G. Ihlen. 2011. *Overvåking av vinterbeiter i Vest-Finnmark og Karasjok 1998–2005–2010*. Resultater fra feltruter. NINA Rapport 745, Trondheim.

Tucker, C., and S. Nicholson. 1999. Variations in the size of the Sahara Desert from 1980 to 1997. *Ambio* 28 (7): 587–591.

Tucker, C., H.E. Dregne, and W.W. Newcomb. 1991. Expansion and contraction of the Sahara desert from 1980 to 1990. *Science* 253: 299–301.

Walters, C.J., and C.S. Holling. 1990. Large-scale management experiments and learning by doing. *Ecology* 71: 2060–2068.

Climate Change, Scarcity and Conflicts in the Sahel

Contents

T. A. Benjaminsen, H. Svarstad, *Political Ecology*,
https://doi.org/10.1007/978-3-030-56036-2_8

Trailer

In February 2008, Erik Solheim visited Mali. He was then the Norwegian Minister of Environment and International Development. Later he became the Executive Director of the United Nations Environment Programme (UNEP) during 2016–2018. As he flew over the arid Sahelian landscape in a light aircraft on his way from Timbuktu to Bamako, he was interviewed by a journalist.

'There are few historical driving forces behind conflicts as strong as environmental degradation and the struggle for resources', said Solheim. 'The struggle for resources was the cornerstone of the warrior culture initiated by Genghis Khan. This merely goes to show how the struggle for resources and environmental degradation are a cause of conflict,' he continued. Journalists had been invited to join the official Norwegian visit to Mali to witness how climate change creates scarcity, which again causes conflicts—in this particular case the Tuareg rebellion that took place in northern Mali between 1990 and 1996.

Overview

Among policy-makers such as Solheim, a narrative of a climate-desertification-conflict nexus holds a strong position, in particular related to discussions about the African Sahel. This narrative resonates also with the international attention to the 'securitization of climate change' (Brown et al. 2007) reflected in policy environments, media, and some academic contributions.

Representing this thinking the journalist Gwynne Dyer's widely acclaimed book on 'climate wars', presents a future scenario where an 'apocalyptic crisis' is 'set to occupy most of the twenty-first century' with 'a probability of wars, if temperatures rise 2 to 3 degrees Celsius' (Dyer 2010: xi-xii). In a similar vein, another popular book also titled 'Climate Wars' carries the subtitle 'Why People Will Be Killed in the twenty-first Century' (Welzer 2012).

In parallel to most contributions to this literature, Welzer's book does not see climate change as the direct cause of conflicts, but more as one of the underlying factors creating declining food production, increased land degradation through more droughts and floods, increased health risks, and more people on the move. Hence, 'the consequences of climate change will reinforce and deepen survival problems and the potential for violence; they will interact with political, economic, ethnic and other social-historical factors and may also lead to open use of force' (Welzer 2012: 75). This thinking is in line with numerous other security analysts writing on this topic and it is frequently referred to as a 'threat multiplier', or a 'catastrophic convergence' resulting from a collision of political, economic, and environmental disasters (Parenti 2011).

The highest international policy levels are in particular attracted to the securitization of climate change. When the UN Security Council debated the links between resource scarcity, development and conflicts in 2007, the Foreign Secretary of the United Kingdom at the time, Margaret Beckett, concluded that 'climate change is transforming the way we think about security', while the French Ambassador to the UN stated that 'the Security Council cannot ignore the threats to international security caused by global warming' (cited by Verhoeven 2014: 785).

In fact, this international attention to the securitization of climate change is found in particular among policy, military and NGO actors, while relatively few researchers advocate this idea (Selby and Hoffman 2014). 'In this, the climate security field diverges sharply from other areas of climate change policy, where scientists have played formative … roles in pushing forward national and international action' (Selby and Hoffmann 2014: 749).

In a historical perspective, this new 'climate reductionism' (giving climate the role as the main variable predicting social change) may be compared to the 'climate determinism' of the nineteenth century and early twentieth century (Hulme 2014). As in the heydays of climate determinism when the agency of colonial subjects was reduced to being a product of African climates, there is also a particular focus on Africa in today's climate reductionism. African drylands, and in particular the Sahel, are usually pointed out as the most prominent example where there is 'a volatile mix of climate change, drought, food shortages, migration and immobility, armed insurrection and heavy weapons proliferation'.[1] The Sahel was also pointed to by the Norwegian Nobel Committee as the prime example of the link between climate change and conflict when former US Vice President Al Gore and the Intergovernmental Panel on Climate Change (IPCC) were awarded the Nobel Peace Prize in 2007.

This narrative about the climate-desertification-conflict nexus in the Sahel consists of two elements. The first says that global climate change leads to drought and desertification, which in turn lead to resource scarcity, while the second holds that this resource scarcity causes migration and the emergence of new conflicts, or it triggers existing, latent conflicts (Benjaminsen 2008, 2016).

The narrative is also deeply Malthusian (Hartmann 2014; Selby and Hoffmann 2014) in its focus on resource scarcity as a cause of environmental degradation, poverty, and an escalating struggle for resources, and with roots in the 'environmental security' school that emerged in the 1990s (Homer-Dixon 1994). Traditionally, this literature has primarily been concerned with 'overpopulation' and the associated 'overuse' of renewable natural resources (Homer-Dixon 1999), but with global warming and the subsequent climate security narrative, the anticipated impact of anthropogenic climate change on the security of societies and livelihoods has gained prominence.

The concept of scarcity as a cause of war and violence has taken root in the media and popular scientific publications too. For instance, in an article published in 1994, the American journalist Robert Kaplan claimed that the conflicts in Liberia, Rwanda and Somalia were unequivocally the result of overpopulation and subsequent environmental crisis in Africa (Kaplan 1994). This article garnered widespread attention and influence, and Tim Worth, then US Undersecretary of

1 Cited from the website of The Center for Climate and Security from a blog on the situation in Mali ► https://climateandsecurity.org/2012/04/23/mali-migration-militias-coups-and-climate-change/

State for Global Affairs, faxed the article to all the US ambassadors in the world (Richards 1996).

Later, a range of journalists and popular-science writers have played a significant role in spreading a Malthusian message to politicians and the general public about conflicts in Africa (e.g. Diamond 2005). The scarcity perspective was also promulgated in a special edition of National Geographic Magazine focusing on Africa in September 2005 (see Moseley 2005 for a critical commentary). Currently, this perspective continues to be repeated in international media and blog posts, which seems to have led to a general strengthening of the climate-desertification-conflict narrative.

In this chapter, we will first present a critical assessment of the literature focusing on scarcity as a main causal factor behind conflicts, which is often referred to as 'the scarcity school' or 'the environmental security school'. Thereafter, we will demonstrate, through a discussion of conflicts in Mali, how the narrative about a climate change-desertification-conflict nexus in the Sahel is misleading and apolitical, since it neglects the context of significant historical and political causes of conflicts.

8.1 Critical Assessment of the Environmental Security School

A major criticism of the environmental security school is that the term 'resource scarcity' is defined so vaguely and broadly that it loses its meaning (Gleditsch 1998; Fairhead 2001; Richards 2005). Since armed conflicts are almost without exception about control over land, such conflicts will necessarily have a resource dimension. However, it does not follow that this dimension explains the conflicts. It is also misleading when such differing processes as environmental degradation, increased population pressure and inequitable access to resources are forced together into a single concept of resource scarcity. In this way, the concept loses its analytical power.

Concepts such as 'needs' and 'scarcity' are not as innocent and neutral as some people would have us believe. For example, on the basis of controversial plans to build a hydropower plant in Sardar Sarovar Narmada in India, Mehta (2007) demonstrates how 'needs' and 'scarcity' are relative and socially constructed concepts that are used purposely to achieve political objectives—in this case modernization through construction of a large hydropower project with enormous social and ecological consequences. In a similar vein, Scoones et al. (2019) point out that scarcity narratives are not merely descriptive storylines, but are constructed, proliferated and adopted with a purpose. They are used actively to legitimize certain policy interventions.

This critique is associated with a general critical assessment within political ecology of the environmental security school. A message from political ecology is that conflicts cannot be understood on the basis of a simple chain of events triggered by resource scarcity, via reduced economic activity and migration, to violent outcomes (Peluso and Watts 2001). Instead, violence is context-specific, at the same time as it is a result of overarching power and relations of production.

Some critics also claim that population growth can also serve to increase the resource base and lead to sustainable agricultural intensification in keeping with the Danish agricultural economist Ester Boserup's theory (Boserup 1965). There are examples from Africa of increases in population combined with favourable government policies that have led to increased investments per unit area in the form of work and capital and thus an improved resource base (e.g. Tiffen et al. 1994; Mortimore 1998; Benjaminsen 2001, see also ► Chap. 9).

The scarcity argument also resonates with predominant Western and Malthusian perceptions of development problems in the Global South. The argument is therefore more a result of Western images of 'the other' than of empirical facts. This 'orientalism' has long Western traditions (Said 1978). In an African context, it entails regarding smallholder farmers and herders as helpless victims and at the same time pathological abusers of landscapes and natural resources. This kind of thinking helps establish external actors such as government bodies and international organizations like the United Nations, in the role of problem solvers (Peluso and Watts 2001).

► Example

Tom Bassett's classic study of the political ecology of farmer-herder conflicts in Côte d'Ivoire presents an example of the complex political and economic causes often found behind such conflicts (Bassett 1988). His main argument is that in order to understand why such conflicts emerge, one needs to study the political economic context behind changes in land-use in individual conflicts, because these conflicts are 'responses in context'. In this particular case, an increased number of Fulani pastoralists from the Sahel came to Côte d'Ivoire after the drought in the mid-1980s, which led to increased crop damage caused by livestock. Farmers responded by confiscating livestock from the herders, which again led to a number of local conflicts—many of which were violent.

These conflicts were interpreted by outside observers as either ethnic clashes or as resource conflicts driven by the combination of a growing population and diminishing resources. While the drought led herders to move south, Bassett shows how the conflicts primarily had external political causes. In order to pay back foreign debt, the Ivorian state established several parastatal companies within the agricultural sector in cooperation with international capital. These companies should, through export, provide foreign exchange to the state. SODEPRA was one such company. Through distributing grazing rights to herders from the Sahel (Mali and Burkina Faso), the aim of increased meat production and meat export should be reached. The influx of pastoralists and livestock to Côte d'Ivoire in the 1980s was therefore a result of an explicit strategy from government to increase the country's meat production to pay its foreign debt. The drought in the Sahel may, however, have made it easier to attract Sahelian pastoralists to Côte d'Ivoire, which indicates that scarcity may potentially play a secondary role in some cases. ◄

The general narrative about scarcity and climate change leading to more violent conflicts has also been extensively criticized in the quantitative literature by for instance Barnett (2003), Nordås and Gleditsch (2007), Salehyan (2008), Buhaug et al. (2014), Buhaug et al. (2015), von Uexkull et al. (2016), and Koubi (2019), although with a minority among quantitative studies arguing for a stronger causal association (Hsiang et al. 2013). In a study based on views among mostly quantitative scholars, Mach et al. (2019) found considerable agreement that state capacity and level of socioeconomic development generally represent the main causes of conflicts. But they also agreed that climate variability and change may play a secondary role in influencing the risk of organized armed conflict within countries.

Generally, the scarcity narrative is logical and it is easy to imagine how scarcity of resources may lead to conflicts. This means that, intuitively, the imagined strong connection between scarcity and conflict makes sense, which can partly explain the power and attraction of this idea. It continues to hold a strong position among policy makers as well as in the media despite most quantitative and qualitative studies questioning such a strong causal connection. It is also interesting to note that Elinor Ostrom received the Nobel Prize in economics in 2009 for three decades of empirical research demonstrating how scarce resources may actually lead to cooperation rather than conflict among resource users under certain conditions (Ostrom 1990, 2005).

Climatic and Environmental Change in the Sahel

There has historically been long term, medium term and short-term climatic variations in the Sahel. During a period from 30,000 to 20,000 B.C.E., the Sahara was humid with large lakes, while in the period 20,000 to 12,000 B.C.E., it was much drier. Thereafter, there was another humid period from around 12,000 to 4000 B.C.E., with lakes in the Sahara and with vegetation that today is found near the Mediterranean, before desiccation gradually took place again (Nicholson 1989).

In addition to these long-term dry and wet periods that have been of importance to how large the desert has been historically, there have also been swings in rainfall during medium-range periods. Nicholson (1989) identified different such dry and wet medium-range periods since the ninth century. In more modern times, both oral tradition and written sources testify to the fact that there have been several drought periods since the seventeenth century (Webb 1995). In other words, droughts have always occurred from time to time in the Sahel, and droughts may be seen as part of the region's climatic system.

Medium-range variations in rainfall can be traced using earlier descriptions of the natural vegetation and of what people planted. After Columbus's journey to America, maize was brought from the 'New World' to Europe and from there to West Africa in the sixteenth and seventeenth centuries. In the beginning of the eighteenth century, it was reported that maize had become the dominant cultivated plant along the Senegal River. However, around 1750, agriculture along this river consisted of a mixture of maize, millet and sorghum, and around 1850, millet and sorghum again were dominant, while maize was almost not planted at all. This shift is explained by the fact that maize needs more rainfall than millet and sorghum. When there is enough rainfall, farmers prefer maize because it produces a better crop. But with less rain, they will with time change to crops that are more drought resistant. Put differently, it was getting gradually drier in the Sahel from the beginning of the eighteenth century to around 1850. Thereafter, there was a humid period until about the turn of the century (Webb 1995).

While the twentieth century was overall relatively dry, the 1950s were unusually wet. According to Descroix et al. (2015), the periods 1900–1950 and 1995–2015 can be considered as periods with average annual rainfall in the last century in the Sahel, while 1951–1967 and 1968–1995 were respectively humid and dry periods. There has since been at least a partial recovery of annual rainfall. Dakar, which is one of the meteorological stations in West Africa with the longest records, demonstrates this trend (◘ Fig. 8.1).

Since it is largely rainfall that drives the Sahelian ecosystem, global warming might in the long run lead to desertification—if it reduces rainfall. However, as demonstrated by Buontempo et al. (2010), there is considerable uncertainty about rainfall trends and projections in the Sahel. There is much less uncertainty related to temperatures, as increasing temperatures have been observed and all models project further increases of surface temperatures in future scenarios (Buontempo et al. 2010).

This uncertainty relating to future rainfall scenarios is underlined by the IPCC in its Fourth and Fifth Assessments (Boko et al. 2007; Niang et al. 2014). In addition, this uncertainty is reflected in the fact that the various climate models do not concur concerning future climate scenarios for the Sahel. While some models support the theory

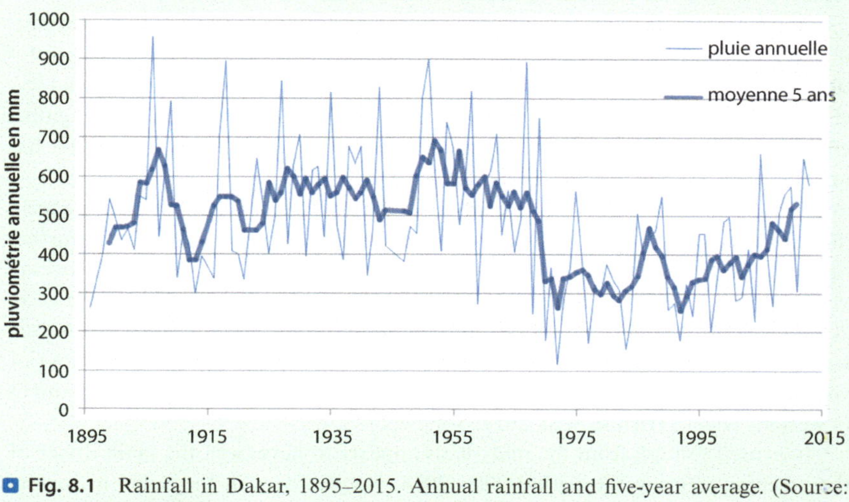

◘ **Fig. 8.1** Rainfall in Dakar, 1895–2015. Annual rainfall and five-year average. (Source: Descroix et al. 2015)

that this region will become drier, other models suggest that there may be more rain in the Sahel in the future. However, Biasutti (2013) finds that most models conclude that the rainy season will be 'more feeble at its start' and 'more abundant at its core'. Hence, the overall trend seems to be towards wetter conditions, but with rainfall more concentrated in time, with more variability, and with higher average temperatures. Of 20 models investigated by Biasutti (2013), only 4 are coming to other conclusions.

Moreover, Giannini (2016) discusses the causes of drought in the Sahel during the late twentieth century reviewing the power of two competing explanations that have been hotly debated. The first posits that human induced land degradation increases the albedo effect from the land surface that again feeds back into the regional climate. The second relies on the influence of the oceans and the monsoon to explain rainfall variation and is therefore more connected to a global climatic system. Giannini (2016: 281–82) concludes that 'drought was caused by large-scale, if subtle, shifts in oceanic temperatures, not by local anthropogenic pressure on the environment' thereby undermining the hypothesis about the albedo effect. Given this conclusion, Giannini also opens up for the possibility that the droughts in the late twentieth century have been influenced by global warming associated with emissions of greenhouse gases.

The idea that deserts are spreading, either due to climate aridification or land-use practices, has, however, existed since early in the colonial period (Davis 2016; Benjaminsen and Hiernaux 2019). In the late nineteenth century and the first decades of the twentieth century, desiccation was understood as the main driver of environmental change in the West African Sahel. However, from the 1930s, the desiccation narrative, or what has been referred to as 'desiccation theory' (Davis 2016) was replaced by a narrative focusing more explicitly on desert advance caused by local mismanagement (Benjaminsen and Hiernaux 2019). Later, the droughts in the Sahel of the 1970s and 1980s enhanced this desertification narrative and led to an institutionalization of desertification at different levels from the regional and national levels in the Sahel to the United Nations, which established desertification as a global environmental problem. This happened despite the greening in the Sahel that has taken place since the droughts in the 1980s.

Already early analyses of the trends after the drought using remote sensing technology indicated a fast recovery of the vegetation cover (Tucker et al. 1991). Further research on better corrected data and longer time series confirmed the overall greening in the Sahel (Olsson et al. 2005; Hutchinson et al. 2005; Herrmann et al. 2005; Fensholt et al. 2006; Dardel et al. 2014; Brandt et al. 2016). There are, however, some local exceptions to the general greening, related either to continued loss of vegetation on shallow soils as in parts of northern Mali or to expanding cropland as in western Niger (Trichon et al. 2017; Tong et al. 2017).

Increased rainfall from the mid-1980s appears to have been the main driver of the greening. During the period 2000–2014, woody cover in the Sahel in general increased by 1.7 ± 5.0%, but with large spatial differences: 0.2 ± 4.2% in densely populated areas versus 2.1 ± 5.2% in sparsely populated areas, 0.9 ± 4.6% in cropped areas versus 2.5 ± 5.4% in savannas and 3.9 ± 7.3% in woodlands (Brandt et al. 2016). Most of these dynamics of the woody population is explained by natural regeneration. However, farmer-induced improvements of parkland management mainly through increased tree planting on farmland as well as some large-scale tree planting in highly eroded landscapes such as in the Ader Doutchi in central Niger may have locally contributed to this greening (Reij et al. 2009).

The overall positive trends observed for both herbaceous and woody plant populations question the main evidence of desertification, which are contrary to the dominating policy narrative on the Sahel represented for instance by the Great Green Wall Initiative, which aims to make the Sahel green and thereby to fight desertification by establishing a 15 km wide and 7100 km long green wall of trees from Senegal to Djibouti. This is an initiative taken by Heads of State in the Sahel and it is funded by the Global Environment Facility at the tune of over 100 million US dollars. In addition, at the climate summit in Paris in December 2015, the project was promised another 4 billion US dollars from donors as well as additional support from France.

Hence, even though international research during the last few decades has undermined the desertification narrative it still lives on in policy initiatives. Climate change may indeed lead to drier conditions and desertification in the long term if rainfall declines. Current climate models predict, however, higher total rainfall in the Sahel following global warming, but also more variable rainfall within a shorter rainy season, in addition to rising temperatures. Therefore, uncertainty about the impacts for people, land-use and ecosystems remains characteristic of climate scenarios for the Sahel.

8.2 Pastoralism, Marginalization and Conflicts in Mali

As we saw in ► Chap. 7, pastoralists are marginalized by government policies in most countries where pastoralism is practised. Pastoral marginalization in Africa is primarily caused by national agricultural policy and legislation related to land tenure.

We will here present two case studies from Mali that illustrate how pastoralists are marginalized in practice and how this can trigger conflicts. The examples also demonstrate some of the more complex historical and political causes behind conflicts. We first discuss Tuareg rebellions in northern Mali before turning to the more recent 'jihadist' uprising in Mali and in the Sahel in general (◘ Fig. 8.2).

Fig. 8.2 Mali. (Source: Benjaminsen et al. 2012)

Tuareg Rebellions in Northern Mali

Tuareg in the Sahel first rebelled against the French colonial invasion from the 1880s until there was a peace treaty in 1916. The French colonial administrator Clauzel, who worked in northern Mali from 1946 until the time of Mali's independence in 1960, wrote a book about his experiences as a colonial official (Clauzel 1989). He argued that the cultural farm heritage of the French resulted in a view of the settled farmer as more productive and connected to the French cause than migrating pastoralists. The French felt that it was especially important to settle pastoralists, not only for purposes of control or ecological considerations, but more generally to promote development.

This imperative was clearly expressed in the Forest Act of 1935. The law introduced two separate types of land: 'land where value has been added' and 'empty' or 'unused' land. Cultivated land was included in the first category, and farmers were given rights to use cultivated land and fallow land that had been used less than 5 years previously. Areas with rich agricultural production in southern Mali were described as 'the useful districts' by the French. Grazing land and forests, on the

Photo 8.1 Tuareg pastoralists in northern Mali have rebelled against the state at several occasions. (Photo: Carsten Sørensen)

other hand, were defined as unused land with no recognized use rights (Benjaminsen and Berge 2004).

For administrative and environmental reasons, the French colonial power drew lines within districts, so that each pastoral group and village was given its own territory where it was required to stay. The French drew up territorial boundaries so that pastoralists were excluded from key resources, with the main goal being to weaken the pastoralists militarily (Photo 8.1).

The boundaries were also meant to keep different groups of pastoralists from coming too close to each other or to villages, in order to prevent conflict and pillage. It was illegal for pastoralists to cross the boundaries without first getting permission from the administration (Berge 2000).

The French colonial power had a mission beyond merely expanding its influence and prestige in the world. It colonized to expand and to civilize. In the colonial archives, there is an oft-repeated expression—*La mission civilisatrice de la France* (the civilizing mission of France). To reach the overall goal of developing and civilizing, France had three goals for its policies in the northern Sahel during the early colonial period; pacifying the population (in other words, ending internal strife and establishing French rule), abolishing slavery, and settling all pastoralists (Berge 2000).

After Mali's independence in 1960, the new government initiated a policy to modernize agriculture. This policy saw the nomadic lifestyle as old-fashioned and unproductive. The enormous grasslands in northern Mali were again referred to as the 'useless' part of Mali (Benjaminsen and Berge 2004).

The Tuareg in northern Mali regarded this modernization policy as a new form of colonialism—only this time the colonizer came from southern Mali leading to further marginalization of the Tuareg. This feeling of increasing marginalization was the main cause of the first Tuareg uprising against the Malian state in 1963 (Ag Baye 1993; Lecocq 2004), which took place during a wet period in the Sahel. The uprising was brutally suppressed by the Malian army using fighter planes and public executions.

After a coup d'état in 1968, the new military government continued its anti-pastoral policy. Pastoralists in northern Mali had little influence on national or local policy, and in practice, the region was governed by a military governor throughout the whole period until a new Tuareg rebellion started in 1990. Many Tuareg experienced this as a period of military occupation (Poulton and Ag Youssouf 1998).

The Tuareg rebellion that took place during 1990–96 resulted in several thousand deaths and a quarter of a million refugees in the neighbouring countries. This is a conflict that advocates of the environmental security school explain through desertification and ensuing scarcity of natural resources. For example, Kahl (2006: 234) claims that in northern Mali, the combination of 'population pressures, poor land use practices, and a fragile ecology ... made soil erosion, desertification, and freshwater scarcity serious problems.'

Kahl also claims that what he calls 'demographic and environmental stress' were important causes of the Tuareg rebellion, without presenting any documentation to back up these claims. It seems that such advocates of the environmental security school have neither studied Sahel's political history nor are familiar with climate or environmental research on the Sahel.

As we have seen, there has been a recovery of both woody and herbaceous vegetation in the Sahel, including in northern Mali, since the drought of the 1980s. The drought played a role in the rebellion of the 1990s, but not in the way the narrative about the correlation between climate change and conflict assumes. Firstly, the droughts led to many Tuareg moving to Algeria and Libya who became politically radicalized, especially by Gaddafi's ideology, which was a mix of Islam and socialism. Many of the Tuareg that went to Libya also ended up as professional soldiers in Gaddafi's army and gained practical experience in warfare in Lebanon and Chad. It was these soldiers that started planning a new uprising in Mali in the 1970s (Lecocq 2004).

Secondly, the droughts led to Mali receiving large amounts of emergency aid. Much of this relief aid intended for northern Mali was allegedly embezzled by corrupt government officials and politicians. In the wealthier parts of Bamako, numerous 'drought palaces' were built in the 1970s and 1980s. These are large houses that many assume were financed by embezzled emergency relief funds (Bourgeot 1990). The rumours of such corruption further stoked the anger of many Tuareg at the Malian authorities (Klute 1995).

There is also increasing competition between Tuareg and settled Songhay farmers over land along the Niger River in northern Mali. The Tuareg, who are primarily pastoralists, are dependent on access to the rich and nutritious *burgu* (*Echinochloa stagnina*) pastures along the river as an important resource in the dry season, while the Songhay grow rice along the river. These two land-uses represent a constant source of small conflicts, a minority of which are violent. However, these types of conflicts had nothing to do with the rebellion, which was started by people with roots in Kidal deep in the Sahara a long way from the river. In fact, the uprising in

1990 came as much of a surprise to the Tuareg further south along the river as it did to everyone else (Poulton and Ag Youssouf 1998).

Thus, the main cause of the Tuareg rebellion that started in 1990 was a modernization policy that led to the marginalization of nomads, combined with an anger at what was perceived as a predatory state. The drought in the mid-1980s led to drier conditions and scarcer resources for farmers and pastoralists alike. However, since the end of the 1980s, there has been a recovery of rainfall in northern Mali. In addition, the first Tuareg uprising in 1963 took place following an unusually wet period (Benjaminsen 2008).

Why Do Pastoralists in Mali Join Jihadist Groups?

In January 2013, Mali again hit international headlines when jihadist forces moved south after having conquered northern Mali in 2012. Five years earlier, the jihadist organization that became *Al-Qaeda in the Islamic Maghreb* (AQIM) had announced its allegiance to al-Qaeda. This group had arrived from Algeria after the military government there had cancelled elections that the Islamists were about to win, which had brought the country to a civil war in the 1990s. This experience further radicalized the Algerian Islamists that survived and escaped to Mali.

In northern Mali, jihadists strengthened their economic position through the kidnapping of foreigners and receiving millions of euros in ransom, and through the involvement in large-scale smuggling of cigarettes and drugs (Daniel 2012; Lounnas 2014; Harmon 2014).

Simultaneously, following NATO's bombing of Libya and the killing of Gaddafi in October 2011, a few thousands Tuareg soldiers in the Libyan army had returned heavily armed to Mali. This sparked a new Tuareg rebellion that had been simmering for a while. However, in 2012, the various jihadist groups established in northern Mali managed to push aside the secular Tuareg rebels, and in January 2013, they started to move south. This led French forces to intervene, and at the battle of Konna in the Mopti region in central Mali, combined French and Malian forces succeeded in pushing back the jihadist troops, and within a few weeks reconquered most of the northern parts of the country.

In 2020, Malian, French, African and UN forces, however, only control the towns in northern and central Mali, while the state and its representatives have withdrawn from the region for security reasons. In the central Mopti region, this means that most of the rural areas have been controlled by various jihadist groups since 2015. These groups have gradually been replacing the state and expanding their authority, support base and influence in the area, because rural populations increasingly relate to and cooperate with the various armed groups.

The main question we want to discuss with this case study is how such increased support for 'jihadist' groups can be explained, thereby to better understand the causes of the current crisis in Mali. It is common in the media and internet blogs to give climate change a considerable role as a driver of the growth and expansion of jihadist groups in the Sahel. Headlines from leading media are for instance: 'How climate change fuels extremism' (CNN, March 2019); 'Drought, expanding deserts and "food for jihad" drive Mali's conflict' (Reuters, April 2015); and 'Climate change is aggravating the conflict in Mali' (BBC, January 2019), while Le Monde claims that 'because of climate change and economic policies, the traditionally difficult situation of nomadic pastoralists in the Sahel is being further degraded to a point where numerous Fulani take up arms to make their claims heard. In majority Muslims, they provide increasing numbers of troops to the jihadist movements that destabilise the region' (our translation from French) (Le Monde, May 2017).

The current scholarly literature on the recent political development in Mali does, however, focus less on climate change and more on how global political economic developments and Islamist thinking and organization relate to national dynamics. This literature has discussed the largely overlapping topics of international jihadism and its tributaries in the country, drug trafficking and hijacking of hostages as sources of funding for armed groups, the dynamics, politics and history of the Tuareg rebellions, the current crisis in the national democratic system and the weakening of the state, the international military intervention and possibilities for peace and stability, the complex and changing Islamic national landscape, and the Malian crisis as a fallout from the Libya conflict.

While many of these contributions are important and give a rich background to the current political crisis in Mali, we argue that a focus on the local political context, and in particular the political ecological context, is necessary in order to more fully explain the expansion of armed groups in the country (Benjaminsen and Ba 2019). This local context is, however, often neglected in the large and rapidly increasing scholarly literature on jihadism in Mali and the Sahel in general.

In order to understand why people join 'jihadist' groups, we argue that it is necessary to understand the political ecology of land and natural resources governance. In the context of land-use conflicts in the central Mopti region, such an approach implies studying peasant motivations in relation to the politics of natural resource access as well as how shifting authority and power relations impact on this access.

In short, the rural peasantry, and especially pastoralists, tend to support jihadist groups, because of an anti-state, anti-elite and pro-pastoral jihadist discourse, because people have become increasingly fatigued and disgruntled by a predatory and corrupt state, which extracts rent from the rural peasantry, and because the development model imposed by the state and international development assistance has not responded to pastoral priorities.

Examples of rent-seeking include land-use conflicts not being resolved, because local officials and judges have received payments from both parties to support their claims, leading to a lack of adjudication, as well as the practices of the Forest Service of for instance randomly fining women collecting fuelwood and herders grazing livestock.

Jihadists leaders generally manage to play on these feelings among pastoralists in particular. The neo-patrimonial state that is the main reason for many of the land-use conflicts in central Mali has in this manner prepared the ground for a gradual jihadist take-over. The result of this growing anti-government sentiment is that many

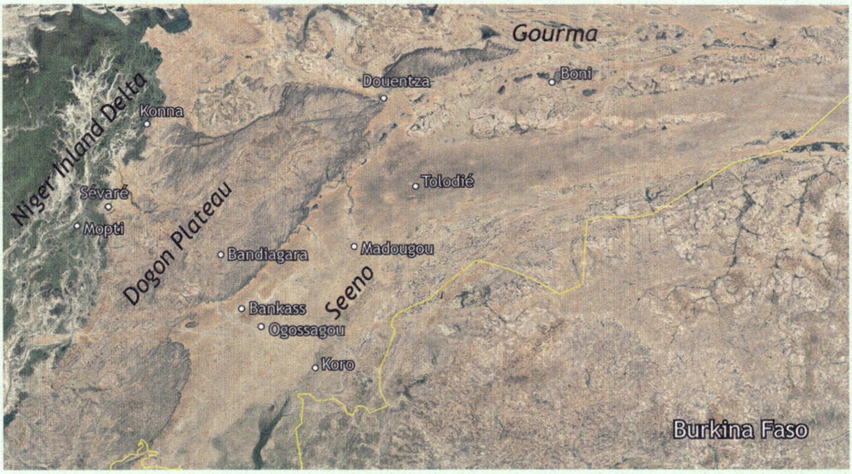

Fig. 8.3 The Seeno plains in Mali where recent conflicts between Dogon and Fulani have taken place. (Source: Edited image from Google Earth)

young men, and in particular pastoralists, have joined various armed groups, and a large part of the rural population tends to see jihadist rule as a lesser evil than the corrupt state.

Recent clashed in central Mali between Dogon farmers and Fulani pastoralists in the dryland Seeno area (Fig. 8.3) may serve as an illustration of the dynamics currently being played out (Benjaminsen and Ba 2019). During spring 2019, these clashes reached international media where they were sometimes presented as 'ethnic clashes' and sometimes as 'climate-driven'. In 2019–2020, hundreds of villages and hamlets in the Seeno were attacked and burned down and more than 1000 people killed.

Especially the attack on Ogossagou village on the 23rd March 2019 caught international media attention when an armed group of Dogon traditional hunters attacked and killed 175 Fulani villagers. About half of the casualties in this horrid attack were children. It immediately made international news headlines and was reported as another example of African 'ethnic violence'. But, since Dogon primarily identify as farmers and Fulani as pastoralists, the violence is also seen as resulting from classic tensions between farmers and herders. Media reports often add that these old conflicts are exacerbated by climate change and population growth leading to increased natural resource scarcity.

From the nineteenth century, Fulani military power dominated the Seeno plains and chased the Dogon in the area back to the nearby escarpment and plateau where most of the Dogon population lived. Those who did not want to abandon their farms had to accept to become integrated into Fulani society as *rimaybe*—Fulani 'slaves' or 'servants'.

In the escarpment and on the plateau above, soils are poor and good farmland is scarce, but with the establishment of French colonial rule in this area from 1905, the Dogon felt safe enough to start the move down again to the plains below to farm (Gallais 1975; Petit 1998). Increasingly, they established cultivation hamlets and opened up new fields (Nijenhuis 2009).

The Seeno had up to Dogon colonization been used seasonally by Fulani pastoralists who see themselves as the owners of pastures under the authority of the Boni Fulani chiefdom that stretches south from Boni town to also include the sandy plains of the Seeno (Van Dijk and De Bruijn 1995). While Dogon since the early twentieth century have transformed vast stretches of pastures to farmland, the new settlements were, however, for decades not seen as problematic by the Fulani as land was thought to be abundant.

This swift colonization has been said to be a result of Dogon 'land hunger' linked to their initial land scarce situation at the escarpment in addition to population growth (Gallais 1975; Petit 1997). The Dogon population increased from about 100,000 in the early twentieth century to at least 300,000 in the early 1970s and about 450,000 around year 2000 (Van Beek 2005). Moreover, the replacement of the hoe by the plough in the 1990s facilitated this expansion, which led to a doubling or even tripling of field sizes (Nijenhuis 2009).

However, with time, Fulani increasingly opposed the expansion of farmland, especially when traditional livestock corridors (*burtol*), which had been established in the nineteenth century, were blocked by new cultivation (De Bruijn and Van Dijk 2005). But since farming has usually been supported by the administration of both the colonial and Malian state at the expense of pastoralism, Fulani pastoralists have seldom been able to prevent pastures from being taken over by cultivation.

After Mali's independence in 1960, the number of Dogon from the escarpment descending to the plains further increased. Moreover, the independent socialist government of Mali followed both an anti-feudal and anti-pastoral policy. This policy encouraged farmers to cultivate new areas without asking traditional authorities for permission.

Land conflicts started to emerge in the 1970s and 1980s. The first violent conflict took place in the pastoral area of Tolodié. In 1975, a large livestock development programme with the World Bank as the main funder, had started to dig a total of 11 wells in the Seeno, including one in Tolodié. The aim of the programme was pastoral development through improved access to water. But by making water permanently available in larger parts of the plains, this water development programme led to a further influx of people to the area. The water rights remained, however, unclear.

While Tolodié had been a pastoral area, the new well also attracted Dogon from the escarpment and plateau who settled to establish hamlets and over a few years transformed the area into an agricultural zone. This led Fulani pastoralists to demand the state administration to accept the creation of a *harima* (a pasture reserve for milk cows that do not follow the annual migration). This was agreed, but not respected by farmers who continued to encroach on these pastures. Over time, a conflict emerged that a local pastoral leader took to court—a case he won in 2002.

However, the administration was bribed by Dogon farmers not to implement the court's decision.

One Dogon even continued to cultivate in the interior of the *harima*, and one day (in August 2002), he was found assassinated. In the night after his body had been discovered, Dogon in the area met, and the next morning they killed the pastoral leader and seven other people.

After this attack, local Fulani mobilized to evict all Dogon from the Tolodié area, which was accepted by the local administration at the time. But in the following years, Dogon managed to gradually move back again to farm in the area.

A few years later, in 2017, some Dogon, without the authorization of the local Dogon village chief, started to demand fees from Fulani herders for grazing around their hamlets. One Dogon also encroached on the *harima* with his fields. On 17 June 2017, when visiting the local market with his son, he was killed by some Fulani (others refer to them as 'jihadists').

This led a group of Dogon to organise an act of revenge, which resulted in the burning down of Fulani hamlets and the killing of 40 Fulani in the Tolodié area. The next day, Fulani burned down Dogon hamlets and killed six Dogon.

This violence in Tolodié in 2002 and 2017 are considered key events leading to the subsequent escalation of Fulani-Dogon violent conflicts. Another key event was the Dogon attack on Sari on 22 May 2012, which is a village situated between Koro and the border to Burkina Faso. In this attack, 350 huts were burnt, 774 cattle taken and 21 Fulani villagers killed in addition to several injured.

This conflict, which is about a cattle corridor being blocked by farming, has also for several years been dealt with by the courts, in this case the primary court in Koro and the appeal court in Sévaré, without any resolution although judgments have been in favour of the Fulani, because they are acknowledged as the first-comers in that area.

The conflict continued with the Dogon associated with and supported by the army, while Fulani pastoralists have joined different jihadist groups in large numbers.

In Mali, militias have often been used by the state in times of crisis to support the armed forces. The army has occasionally acknowledged that it has provided training and logistical support to these militias (Boisvert 2015). According to Charbonneau (2020), the French Barkhane force also works with various pro-government militias in their anti-terrorism efforts in Mali. This has provided the various militias with increased power and further exacerbated inter-community tensions.

Dana Amassagou is the latest addition to pro-government militias in Mali. It is usually referred to as a Dogon militia consisting of traditional Dogon hunters as well as former army officers and soldiers of Dogon ethnicity, but also mixed with some combatants who are of Bambara ethnicity.

This militia was established in December 2016 to protect the Dogon population against jihadist attacks. In 2018, *Dana Amassagou* experienced a boost in its operational power when it started to receive support from the state after an initiative from the prime minister at the time. The support included the provision of automatic hand weapons (e.g. Kalashnikovs) and RPG grenade launchers. It also allowed *Dana Amassagou* to set up several training camps.

The aim of *Dana Amassagou* also seems to have changed over time from defending Dogon villages to actively attacking Fulani villages. In 2018, the militia reportedly started to systematically attack Fulani villages through setting fire to buildings and chasing the civilian Fulani away. Since Fulani are widely regarded as associated with jihadism, this operation may be seen as a state-supported form of counterinsurgency. Several thousand Fulani have, as a result, crossed the border to Burkina Faso to seek refuge.

Hence, while jihadists generally kill individuals who oppose them or who cooperate with the army or state administration, but do not attack the civilian population in general, the strategy of the pro-government militia 'is not to confront the jihadists who are responsible for the attack but rather to aim at soft targets—in this case, Peul (Fulani) civilians. Also, sometimes armed groups take advantage of the context of conflicts to settle old accounts and disputes, including conflicts over natural resources' (Ibrahim 2017: 18).

The active state support to *Dana Amassagou* continued until April 2019 when public criticism and demonstrations in Bamako led former President Keita to appoint a new prime minister.

Hence, to sum up this case study, what drives pastoralists to join 'jihadist' groups is not primarily religious conviction. It is to end the domination by traditional elites in collusion with the state administration, and to resist the army's violence and human rights abuse against pastoral communities. There are numerous reports that the army has arrested and even executed many Fulani men accused of being 'jihadists'.

Farmer-herder conflicts in the Sahel tend to be presented by external observers as either driven by ethnicity or by increasing resource scarcity caused by climate change or population growth—or a combination of these factors. While these elements may be relevant and play a role, which we also see in the case of Fulani-Dogon violence in Mali, narratives about ethnicity or scarcity also serve to simplify explanations of conflicts and to gloss over their more complex root causes linked to historical and contemporary land relations.

We therefore argue that in order to more fully understand such land struggles that escalate into violence, it is necessary to study the material politics and political ecology of land governance in a historical context. This will lead to explanations that move beyond simple causal chains where the Sahel is seen as an arena with decreasing resource scarcity affecting livelihoods that again trigger migrations or conflicts.

In addition to demonstrating the complex background to the current crisis in Mali and how simplistic narratives about its causes are not helpful for understanding this crisis, the case also shows how views in a military intervention of the enemy as 'terrorists' or 'jihadists' are also dangerous and able to further fuel violent conflicts.

Conclusion

The causes of the conflicts in Mali and in the Sahel in general are primarily historical and political relating to state policies leading to marginalization of pastoralists. Climate change and resource scarcity only play a minor role in explaining these conflicts.

This is one of several examples from the field of 'environment and development' where empirical research challenges dominating narratives without this appearing to have any influence on policies. Roe (1994) claims that policy narratives often resist change, and even in the face of data that contradict the narrative being presented, because these narratives support the assumptions of decision-makers in a world marked by uncertainty, complexity and polarization. Narratives provide a simple frame of reference that it is easy to relate to. It is much harder to relate to uncertainty and complexity. The narrative about climate-desertification-conflict in the Sahel is also a moral narrative that helps establish a clear link between the West's responsibility for the greenhouse effect on the one hand and poverty and war in Africa on the other. Although, on an individual level, it is easy to sympathise with this kind of agenda, political ecologists should also critically investigate these types of 'moral' narratives to check whether they are empirically tenable.

In addition, while this narrative implies no personal costs to its proponents (or their constituencies), it provides an image of being environmentally concerned 'free of charge' so to speak.

Some narratives are also especially attractive to people in power, politicians and bureaucrats, simply because the basic elements of the narrative or its assumptions support the careers of powerful institutions or individuals (Leach and Mearns 1996). These narratives then gain momentum and present irrefutable facts, regardless of what empirical research tells us. Since they are so closely intertwined with the interests of powerful institutions and individuals, the narratives are also very hard to change. In this way, the narrative about climate-desertification-conflict in the Sahel is also one of many examples showing that better research does not necessarily yield better policies.

Questions

1. How do you see the role of resource scarcity as a cause of violent conflicts?
2. How is climate change impacting on the Sahel?
3. Why do pastoralists join jihadist groups in Mali?
4. Why are policy narratives difficult to change?

References

Ag Baye, C. 1993. The process of a peace agreement. Between the movements and the United Fronts of Azawad and the government of Mali. In '…*Never drink from the same cup'. Proceedings of the conference on indigenous peoples in Africa. Tune, Denmark, 1993*. IWGIA Document No. 74, ed. H. Veber, J. Dahl, F. Wilson, and E. Wæhle, 247–256. Copenhagen: IWGIA.

Barnett, J. 2003. Security and climate change. *Global Environmental Change* 13: 7–17.

Bassett, T.J. 1988. The political ecology of peasant-herder conflicts in the northern Ivory Coast. *Annals of the Association of American Geographers* 78 (3): 453–472.

Benjaminsen, T.A. 2001. The population-agriculture-environment nexus in the Malian cotton zone. *Global Environmental Change* 11 (4): 27–39.

———. 2008. Does supply-induced scarcity drive violent conflicts in the African Sahel? The case of the Tuareg rebellion in northern Mali. *Journal of Peace Research* 45 (6): 831–848.

———. 2016. Does climate change lead to conflicts in the Sahel? In *The end of desertification? Disputing environmental change in the drylands*, ed. R. Behnke and M. Mortimore. Heidelberg: Springer.

Benjaminsen, T.A., and B. Ba. 2019. Why do pastoralists in Mali join jihadist groups? A political ecological explanation. *Journal of Peasant Studies* 46 (1): 1–20.

Benjaminsen, T.A., and G. Berge. 2004. Myths of Timbuktu – From African El Dorado to desertification. *International Journal of Political Economy* 34 (1): 31–59.

8 Benjaminsen, T.A., and P. Hiernaux. 2019. From desiccation to global climate change: A history of the desertification narrative in the West African Sahel, 1900-2018. *Global Environment* 12: 206–236.

Benjaminsen, T.A., K. Alinon, H. Buhaug and J.T. Buseth. 2012. Does climate change drive land-use conflicts in the Sahel? *Journal of Peace Research* 49 (1), 97–111.

Berge, G. 2000. *In defence of pastoralism: Form and flux among Tuaregs in northern Mali*. PhD dissertation, Department of Social Anthropology, University of Oslo.

Biasutti, M. 2013. Forced Sahel rainfall trends in the CMIP5 archive. *Journal of Geophysical Research: Atmospheres* 118: 1613–1623.

Boisvert, M.A. 2015. Failing at violence: The longer-lasting impact of pro-government militias in northern Mali since 2012. *African Security* 8: 272–298.

Boko, M., I. Niang, A. Nyong, C. Vogel, A. Githeko, M. Medany, B. Osman-Elasha, R. Tabo, and P. Yanda. 2007. Africa. In *Climate change 2007: Impacts, adaptation and vulnerability*, ed. M. Parry, O. Canziani, J. Palutikof, P. van der Linden, and C. Hanson, 433–467. Cambridge: Cambridge University Press.

Boserup, E. 1965 (1993). *The conditions of agricultural growth*. London: Earthscan.

Bourgeot, A. 1990. Identité touarègue: de l'aristocratie à la revolution. *Etudes Rurales* 120: 129–162.

Brandt, M., P. Hiernaux, K. Rasmussen, C. Mbow, L. Kergoat, T. Tagesson, Y.Z. Ibrahim, A. Wélé, C.J. Tucker, and R. Fensholt. 2016. Assessing woody vegetation trends in Sahelian drylands using MODIS based seasonal metrics. *Remote Sensing of Environment* 183: 215–225.

Brown, O., A. Hammil, and R. McLeman. 2007. Climate change as the 'new' security threat: Implications for Africa. *International Affairs* 83 (6): 1141–1154.

Buhaug, H., J. Nordkvelle, T. Bernauer, T. Böhmelt, M. Brzoska, J.W. Busby, A. Ciccone, H. Fjelde, E. Gartzke, N.P. Gleditsch, J.A. Goldstone, H. Hegre, H. Holtermann, V. Koubi, J.S.A. Link, P. Lujala, J. O'Loughlin, C. Raleigh, J. Scheffran, J. Schilling, T.G. Smith, O.M. Theisen, R.S.J. Tol, H. Urdal, and N. Von Uexcull. 2014. One effect to rule them all? A comment on climate and conflict. *Climate Change* 127 (3–4): 391–397.

Buhaug, H., T.A. Benjaminsen, E. Sjaastad, and O.M. Theisen. 2015. Climate variability, food production shocks, and violent conflict in sub-Saharan Africa. *Environmental Research Letters* 10: 125015.

Buontempo, C., B. Booth, and W. Moufouma-Okia. 2010. The climate of the Sahel. In *Global security risks and West Africa. Development challenges*, ed. P. Heinrigs and M. Trémolières, 58–71. Paris: Organisation for Economic Co-operation and Development.

Charbonneau, B. 2020. Sahel: la gouvernance contre-insurrectionnelle. *Bulletin Francopaix* 5 (1): 1–8.

Clauzel, J. 1989. *Administrateur de la France d'Outre-Mer*. Paris: Editions Jeanne Laffitte.

Daniel, S. 2012. *AQMI. L'industrie de l'enlèvement*. Mesnil-sur-l'Estrée: Fayard.

Dardel, C., L. Kergoat, E. Mougin, P. Hiernaux, M. Grippa, and C.J. Tucker. 2014. Re-greening Sahel: 30 years of remote sensing data and field observations (Mali, Niger). *Remote Sensing of Environment* 140: 350–364.

Davis, D.K. 2016. *The arid lands: History, power, knowledge*. Cambridge: MIT Press.

De Bruijn, M., and H. Van Dijk. 2005. Moving people: Pathways of Fulbe pastoralists in the Hayre-Seeno area, Central Mali. In *Sahelian pathways: Climate and society in central and south Mali*, Report 78/2005, ed. M. De Bruijn, H. Van Dijk, M. Kaag, and K. Van Til, 247–273. Leiden: African Studies Centre.

Descroix, L., A. Diongue Niang, G. Panthou, A. Bodian, Y. Sane, H. Dacosta, M. Malam Abdou, J.P. Vandervaere, and G. Quantin. 2015. Évolution récente de la pluviométrie en Afrique de l'ouest à travers deux régions: la Sénégambie et le Bassin du Niger Moyen. *Climatologie* 12: 25–43.

Diamond, J. 2005. *Collapse: How societies choose to fail or survive*. London: Allen Lane.

Dyer, G. 2010. *Climate wars. The fight for survival as the world overheats*. Oxford: Oneworld.

Fairhead, J. 2001. International dimensions of conflict over natural and environmental resources. In *Violent environments*, ed. N.L. Peluso and M. Watts, 213–236. Ithaca and London: Cornell University Press.

Fensholt, R., I. Sandholt, S. Stisen, and C. Tucker. 2006. Analysing NDVI for the African continent using the geostationary meteosat second generation SEVIRI sensor. *Remote Sensing of Environment* 101 (2): 212–229.

Gallais, J. 1975. *Pasteurs et paysans du Gourma: La condition Sahelienne*. Paris: Centre National de la Recherche Scientifique.

Giannini, A. 2016. 40 years of climate modelling: The causes of late-20th century drought in the Sahel. In *The end of desertification? Disputing environmental change in the drylands*, ed. R. Behnke and M. Mortimore. Heidelberg: Springer.

Gleditsch, N.P. 1998. Armed conflict and the environment: A critique of the literature. *Journal of Peace Research* 35 (3): 363–380.

Harmon, S.A. 2014. *Terror and insurgency in the Sahara-Sahel region*. London: Routledge.

Hartmann, B. 2014. Converging on disaster: Climate security and the Malthusian anticipatory regime for Africa. *Geopolitics* 19: 757–783.

Herrmann, S.M., A. Anyamba, and C.J. Tucker. 2005. Recent trends in vegetation dynamics in the African Sahel and their relationship to climate. *Global Environmental Change* 15 (4): 394–404.

Homer-Dixon, T. 1994. Environmental scarcities and violent conflict. Evidence from cases. *International Security* 19 (1): 5–40.

———. 1999. *Environment, scarcity, and violence*. Princeton: Princeton University Press.

Hsiang, S., M. Burke, and E. Miguel. 2013. Quantifying the influence of climate on human conflict. *Science* 341 (6151): 1235367.

Hulme, M. 2014. Reducing the future to climate: A story of climate determinism and reductionism. *Osiris* 26 (1): 245–266.

Hutchinson, C.F., S.M. Herrmann, T. Maukonen, and J. Weber. 2005. Introduction: The 'greening' of the Sahel. *Journal of Arid Environments* 63: 535–537.

Ibrahim, I.Y. 2017. *The wave of jihadist insurgency in West Africa: Global ideology, local context, individual motivations*, West African Papers, No. 07. Paris: OECD Publishing.

Kahl, C.H. 2006. *States, scarcity, and civil strife in the developing world*. Princeton: Princeton University Press.

Kaplan, R.D. 1994. The coming anarchy: How scarcity, crime, overpopulation, and disease are rapidly destroying the social fabric of our planet. *Atlantic Monthly*, February, pp. 44–76.

Klute, G. 1995. Hostilités et alliances. Archéologie de la dissidence des touaregs au Mali. *Cahiers d'Études Africaines* 137 (XXXV-1): 55–71.

Koubi, V. 2019. Climate change and conflict. *Annual Review of Political Science* 22: 343–360.

Leach, M., and R. Mearns, eds. 1996. *The lie of the land: Challenging received wisdom on the African environment*. Oxford: James Currey.

Lecocq, B. 2004. Unemployed intellectuals in the Sahara: The *teshumara* nationalist movement and the revolutions in Tuareg society. *International Review of Social History* 49 (Supplement 12): 87–109.

Lounnas, D. 2014. Confronting Al-Qa'ida in the Islamic Maghrib in the Sahel: Algeria and the Malian crisis. *The Journal of North African Studies* 19 (5): 810–827.

Mach, K.J., C.M. Kraan, W.N. Adger, H. Buhaug, M. Burke, J.D. Fearon, C.B. Field, C.S. Hendrix, J.-F. Maystadt, J. O'Loughlin, P. Roessler, J. Scheffran, K.A. Schultz, and N. Von Uexkull. 2019. Climate as a risk for armed conflict. *Nature* 571: 193–197.

Mehta, L. 2007. Whose scarcity? Whose property? The case of water in western India. *Land Use Policy* 24 (4): 654–663.

Mortimore, M. 1998. *Roots in the African dust: Sustaining the sub-Saharan drylands*. Cambridge: Cambridge University Press.

Moseley, W.G. 2005. Reflecting on National Geographic Magazine and academic geography: The September 2005 special issue on Africa. *African Geographical Review* 24: 93–100.

Niang, I., O.C. Ruppel, M.A. Abdrabo, A. Essel, C. Lennard, J. Padgham, and P. Urquhart. 2014. Africa. In *Climate change 2014: Impacts, adaptation, and vulnerability. Part A: Global and sectoral aspects*, ed. C.B. Field, V.R. Barros, D.J. Dokken, K.J. Mach, and M.D. Mastrandrea, 1199–1265. Cambridge: Cambridge University Press.

Nicholson, S. 1989. Long-term changes in African rainfall. *Weather* 44 (2): 46–56.

Nijenhuis, K. 2009. Reconsidering conflicts over land in the Sahel as conflicts over power. In *Legal anthropology from the low countries*, ed. A. Böcker, W. Van Rossum, and H. Weyers, 69–101. Amsterdam: Reed Business.

Nordås, R., and N.P. Gleditsch. 2007. Climate change and conflict. *Political Geography* 26: 627–638.

Olsson, L., L. Eklundh, and J. Ardö. 2005. A recent greening of the Sahel–trends, patterns and potential causes. *Journal of Arid Environments* 63: 556–566.

Ostrom, E. 1990. *Governing the commons: The evolution of institutions for collective action*. Cambridge: Cambridge University Press.

———. 2005. *Understanding institutional diversity*. Princeton: Princeton University Press.

Parenti, C. 2011. *Tropic of chaos. Climate change and the new geography of violence*. New York: Nation Books.

Peluso, N.L., and M. Watts, eds. 2001. *Violent environments*. Ithaca and London: Cornell University Press.

Petit, V. 1997. Société d'origine et logiques migratoires. Les Dogon de Sangha (Mali). *Population* 52 (3): 515–543.

———. 1998. *Les migrations Dogon*. Paris: L'Harmattan.

Poulton, R.E., and I. Ag Youssouf. 1998. *A peace of Timbuktu. Democratic governance, development and African peacemaking*. New York and Geneva: UNIDIR.

Reij, C., G. Tappan, and M. Smale. 2009. Re-greening the Sahel: Farmer-led innovation in Burkina Faso and Niger. In *Millions fed: Proven successes in agricultural development*, ed. D.J. Spielman and R. Pandya-Lorch. Washington, DC: IFPRI.

Richards, P. 1996. Fighting for the rain forest: war, youth and resources in Sierra Leone. Oxford: James Currey.

Richards, P., ed. 2005. *No peace, no war: An anthropology of contemporary armed conflicts*. Oxford: James Currey.

Roe, E. 1994. *Narrative policy analysis: Theory and practice*. Durham: Duke University Press.

Said, E. 1978. *Orientalism. Western conceptions of the orient*. London: Penguin.

Salehyan, I. 2008. From climate change to conflict? No consensus yet. *Journal of Peace Research* 45 (3): 315–326.

Scoones, I., R. Smalley, R. Hall, and D. Tsikata. 2019. Narratives of scarcity: Framing the global land rush. *Geoforum* 101: 231–241.

Selby, J., and C. Hoffman. 2014. Rethinking climate change, conflict and security. *Geopolitics* 19: 747–756.

Tiffen, M., M. Mortimore, and F. Gichuki. 1994. *More people, less erosion. Environmental recovery in Kenya*. Chichester: John Wiley & Sons Ltd.

Tong, X., M. Brandt, P. Hiernaux, S.M. Herrmann, F. Tian, A.V. Prishchepov, and R. Fensholt. 2017. Revisiting the coupling between NDVI trends and cropland changes in the Sahel drylands: A case study in western Niger. *Remote Sensing of Environment* 191: 286–296.

Trichon, V., P. Hiernaux, R. Walcker, and E. Mougin. 2017. The persistent decline of patterned woody vegetation: The tiger bush in the context of the regional Sahel greening trend. *Global Change Biology* 2018: 1–16.

Tucker, C., H.E. Dregne, and W.W. Newcomb. 1991. Expansion and contraction of the Sahara desert from 1980 to 1990. *Science* 253: 299–301.

Van Beek, W. 2005. The Dogon heartland: Rural transformations on the Bandiagara escarpment. In *Sahelian pathways: Climate and society in central and south Mali*, Report 78/2005, ed. M. De Bruijn, H. Van Dijk, M. Kaag, and K. Van Til, 40–70. Leiden: African Studies Centre.

Van Dijk, H., and M. De Bruijn. 1995. Pastoralists, chiefs and bureaucrats: A grazing scheme in dryland central Mali. In *Local resource management in Africa*, ed. J.P.M. van den Breemer, C.A. Drijver, and L.B. Venema, 77–95. Chichester: John Wiley & Sons.

Verhoeven, H. 2014. Gardens of Eden or hearts of darkness? The genealogy of discourses on environmental insecurity and climate wars in Africa. *Geopolitics* 19: 784–805.

Von Uexkull, N., M. Croicu, H. Fjelde, and H. Buhaug. 2016. Civil conflict sensitivity to growing-season drought. *PNAS* 113 (44): 12391–12396.

Webb, J. 1995. *Desert frontier: Ecological and economic change along the Western Sahel, 1600–1850*. Madison: University of Wisconsin Press.

Welzer, H. 2012. *Climate wars. Why people will be killed in the 21st century*. Cambridge: Polity.

Population Growth, Markets and Sustainable Land-Use in Africa

Contents

T. A. Benjaminsen, H. Svarstad, *Political Ecology*,
https://doi.org/10.1007/978-3-030-56036-2_9

Trailer

Is a growing population an obstacle to agricultural development in Africa? To what extent does population growth lead to environmental degradation? Does increased market integration in agriculture promote or undermine sustainability? These are some key questions that we will discuss in this chapter and which have been debated in the development literature for several decades.

Overview

Discussions about population growth and the role of the market are also important in political ecology and can be traced back to the theories of Malthus and Marx. From the emergence of the field in the 1980s, political ecology has been influenced by Marxist and Neo-Marxist theories. But, as political ecology also became subject to influence from actor-oriented perspectives (e.g. Norman Long, Elinor Ostrom, James Scott) and poststructuralism (e.g. Michel Foucault, Edward Said, Maarten Hajer), internal tensions and discussions emerged. Simply summarized, one may say that the criticism of Malthus has persisted, while the Marxist perspective has been nuanced and further developed over time.

This chapter sets out to illustrate these discussions through the presentation of case studies of agricultural development and its environmental and livelihood consequences in Africa. The chapter shows how increased agricultural production in areas that lack capital is dependent on a certain population density, how access to markets is necessary for agricultural development, how market-oriented agriculture need not necessarily result in environmental degradation, but also how, under certain policy conditions, international capital may undermine the livelihoods of smallholders.

9.1 Population Growth and Agricultural Development in Africa

In policy discussions about African agriculture, terms such as 'overpopulation', 'underdevelopment' and 'environmental crisis' are taken for granted and used with great certainty. While human fertility rates have generally been declining in developing countries during the last 40 years, population growth still remains high in Sub-Saharan Africa. This growth is due to a gap between mortality rates that have been declining for decades and fertility rates that only since the 1990s begun to show a considerable decline. The relatively large decline in the number of deaths compared with the number of births explains why Africa's population growth is the highest in the world. Fertility, that is, the number of children a woman on average gives birth to, was 6.7 for Africa overall at the beginning of the 1970s, while in 2017 it had come down to 4.7. According to the UN, it will drop down to 3.1 in 2050 (UN 2015). In other words, there will be high population growth in Africa for many more years. The question is what consequences this will have for agricultural development on the continent.

While grain production per capita in Asia and Latin America has increased significantly in the last 50 years, it has fallen in Africa, at least according to official statistics. These statistics show for instance a decline of 15% in agricultural production per capita from 1969 to 2004 in Africa south of the Sahara (Wiggins 2005).

There are, however, signs that this negative trend in African agricultural development may have changed towards a more positive trajectory since the early 2000s (Wiggins 2014). Today, there is more optimism about African agriculture among governments, foreign aid donors and researchers than there has been for many decades. Hence, the pessimism of the 1970s, 1980s and 1990s has been replaced by an optimism about the development potential of African agriculture. While growth in food production in the 1970s and 1980s was below population growth, from 1990 to 2010 it was 3.2% per year for Africa in general, which is just above population growth for this period (Wiggins 2014). These are, however, official numbers from FAO that to some extent are subject to negotiations and internal politics in the ministries of agriculture in the various African countries (Jerven 2013). This may lead to both over-reporting to please superiors and politicians as well as under-reporting of certain crops that are subject to less political attention. In spite of this uncertainty, there are signs that agricultural production has increased over the last decade or two. Also, economic growth has picked up on the continent. For instance, according to Radelet (2010) the economies of 17 countries in Africa grew on average by 5% per year between 1996 and 2008, compared to no growth during 1975–1995.

Despite these positive signs, small-scale farming in Africa is still considered by governments and international aid donors as an obstacle to modernization and progress. As noted by Scoones et al. (2014: 14): 'African agriculture is often depicted as stagnant, underproductive, and a cause of land degradation, in need of revival through integration with large-scale, commercial operations'. Hence, the assumption is that a 'take off' in African agricultural development needs to be mobilized from the outside.

Definition

The triple F crisis (finance, food, fuel) in 2007–2008 caused international capital to 'discover' land in Africa as an investment object, which intensified international debates about *land grabbing*.

While the narrative about Africa as overpopulated has dominated development debates for decades, with the triple F crisis a new narrative about large unused land areas waiting for development and investments emerged in parallel. Often these two narratives are currently mixed into one, which says that African smallholders tend to farm in economically inefficient ways that also lead to soil degradation and loss of soil nutrients. Therefore, more modern and technologically advanced forms of farming should take over. In practice, this often means that foreign capital displaces smallholders from areas they have historically used. This is made possible because small-scale farmers and pastoralists usually are not granted formal land rights from the state. Their customary and historical use of land is often not sufficiently recognized in national legislation, which tends to give the state overall rights to most land.

In this chapter, we give examples of such land grabbing. This is a metaphor that has been successfully used by a number of journalists, activists and scholars to highlight unequal power relations and that some powerful actors manage to 'grab' land from smallholders who have historically used the land. Such land grabbing is the result of modernization thinking that sees development as a linear progression from small-scale farming to industrialized large-scale agriculture linked to global markets (McKeon 2015; Bergius et al. 2018). Some entrepreneurial smallholders are expected to integrate into these global value chains, while others who are unable to switch from a 'peasant mentality' to become modern farmers are expected to leave agricultural production altogether (Bergius et al. 2018). In fact, the transfer of land to more efficient farmers ('land mobility') lies at the heart of this modernization thinking. Its implication is that the scale of agriculture is increased, while the number of people making a living from agriculture is reduced (Woodhouse 2012; McKeon 2015; Bergius et al. 2018). This process again leads to de-peasantization as smallholders are squeezed out and forced to migrate to less fertile lands or urban slums (Araghi 1995, 2000; Davis 2006).

However, we argue in this chapter that contrary to the assumptions behind modernization thinking, small-scale farming may perform well both economically and environmentally, and not the least employ a maximum number of people if facilitated by government policies.

► Example

The Machakos District in Kenya represents a well-known case in the political ecology and development studies literature that supports the idea that smallholder production may be sustainable (Tiffen et al. 1994). In the 1930s, this district was marked by soil erosion and what colonial environmental and agricultural officials saw as land degradation caused by overpopulation. While the population in the district was 240,000 in 1932, it had increased to 1.4 million in 1990. At the same time, erosion had, however, been reduced and the environment improved through widespread tree planting and terracing leading to large increases in production of coffee and maize both per capita and per area. The researchers behind this study concluded that this development was possible due to population increase providing more manpower to work the land combined with a policy change in Kenya after independence. During colonization, coffee production for the market had been reserved for white settlers, but with independence, small-scale producers got access to the coffee market in nearby Nairobi. Hence, population increase combined with access to coffee as a cash-crop facilitated this development that has been labelled 'sustainable intensification' (Tiffen et al. 1994). ◄

A recent longitudinal case study from central Tanzania for the period 1991–2016 also found substantially increased income and improved livelihoods generated by small-scale cash-crop farming, in this case farming of sunflower (Östberg et al. 2018). In a related study from the same research group from rural northern Tanzania, Brockington et al. (2018) also found significantly increased prosperity levels among small-scale farmers from the early 1990s to 2013. In a similar vein, Snyder et al. (2020) show how Tanzanian small-scale farmers responded swiftly to changes in the political economic context, which 'debunks the idea that rural farmers are slow to respond to "modern" farming methods or that smallholder farming is stagnant and cannot reduce poverty' (p. 33).

According to Wiggins (2005) agricultural growth areas normally are densely populated and with proximity to markets, while the areas where there is less agricultural growth are usually sparsely populated, marginal regions with poor access to markets. Studies of success stories of agricultural development in Africa show two common characteristics: firstly, a correlation between high population density and agricultural growth, and secondly, that this growth is also largely driven by access to markets (Wiggins 2005).

The close relationship between population density and growth in agriculture that has been observed from case studies in Africa supports the theory of agricultural development introduced by the Danish agricultural economist, Ester Boserup (1910–1999) (Boserup 1965). This theory was an explicit attack on Malthusian views on the impact of population growth on agricultural production. Boserup did not view population growth as a problem, as Malthus did, but rather as a resource. In her opinion, a certain population density is a prerequisite for technological innovation and productivity improvements in agriculture. She also considered population growth as a factor that fosters agricultural intensification through technological change and innovation.

Intensification can also at one time or the other result in increased labour investments in upgrading the land. This may be in the form of tree planting and construction of dikes, terraces, dams and contour lines that will increase the resource base. Thus, carrying capacity is not taken as a fixed entity, but is seen as flexible.

Hence, a certain population density may be a necessary condition for increasing agricultural production in areas where first and foremost it is the labour of the people that drives development—in the absence of investment capital. As a second factor, there is the need to access markets to sell surplus production.

In the next sections, we will first present a case from northern Mali, which may be said to be an area that is under-populated for agricultural development. Thereafter, we shall introduce research from the cotton zone in southern Mali, which shows the market's role as an engine for agricultural development. This case also discusses possible environmental consequences of intensification in agriculture. Finally, we present an example of how capital and the market can play opposite and even destructive roles in relation to agricultural development in Africa.

Labour Constraints in Northern Mali

As a response to recurrent droughts in the Sahel, there have been attempts to make nomadic pastoralists diversify their mode of subsistence (Pedersen and Benjaminsen 2008). National governments and international development agencies often claim that it is important for pastoralists not to depend only on one source of income. Agriculture is therefore frequently promoted in areas that are dominated by animal husbandry. Ensuing diversification projects have often been carried out in conjunction with attempts at sedentarizing pastoralists.

In addition to old and mainly politically motivated arguments for governments to settle nomads and to make them diversify their activities in areas such as the northern Sahel, the focus on livelihoods diversification in the development literature also tends to provide arguments for diversification policies. Within the 'sustainable livelihoods framework' the work of Ellis (2000) has for instance been taken to support this view.

Economic diversification typically implies higher demands on household labour. For instance, studying Fulani groups in Burkina Faso, Hampshire and Randall (2005) found that a natural response to such labour demands has been the maintenance of high fertility rates making it possible for larger households to diversify more easily.

While increased diversification of livelihoods might be a good idea for many rural African households, the question is under what conditions it is a good advice to follow, and whether it is such a good idea for the more mobile pastoralists in the driest parts of the continent. Let us present one such example.

In the Timbuktu region of northern Mali, the Niger River flows through an arid region south of the Sahara desert. The river represents a rich resource in an otherwise barren landscape, where herding of livestock dominates. In the rainy season, the water of the Niger spreads out over large plains, where rice has been cultivated for thousands of years (Ki-Zerbo 1978: 56). This is a sparsely populated area and the methods of rice cultivation are still relatively rudimentary. Because of lack of manpower and capital, farmers can only build simple levees and floodgates to attempt to check the water level. The system is therefore fragile, and it totally depends on the water flowing smoothly and gradually without drowning the floating rice. Often, the water flows quickly onto the rice fields or the levees burst, so that the water disappears from the rice fields and the rice plants dry out. It is perhaps only in one of three years that this traditional system provides any substantial crops. Several aid organizations, such as the Norwegian Church Aid (NCA), have attempted to improve the system by investing in more robust levees and floodgates for better control of the floodwater. These investments have to an extent reduced, but not eliminated, the risk (◘ Photo 9.1).

Low population density has for a long time remained one of the biggest obstacles to agricultural development along this stretch of the Niger river. The labour shortage is particularly evident where Tuareg pastoralists control the land. These are places along the river where Tuareg have been encouraged to settle and cultivate rice. As long as the NCA paid people through a 'food for work' programme during the 1980s, labour was available to build levees and floodgates and to stay and cultivate rice. But as soon as this support was discontinued in the 1990s, pastoralists had the choice between resuming their nomadic life or staying in one place to farm the land. The latter entailed investing their labour in a very risky and vulnerable agricultural production. Usually, they chose to resume their nomadic life.

A concrete example of this is the hamlet called Banguel, which lies a few miles west of Gourma Rharous—the administrative centre of the district. The Tuaregs based in Banguel are from the group *Igouadaren Ouest*. This group consists of some

Photo 9.1 Improved floodgates on the Niger river in northern Mali funded by the Norwegian Church Aid. (Photo: Norwegian Church Aid)

families with high status warriors (*imoushar*), some craftsmen (*inhaden*) and a large majority of former slaves, called *bella* (or *eklan*). The bella make up the workforce and are still loyal to the *imoushar*, who take all the decisions. The group controls two large rice fields—one of 97 hectares and another of 248 hectares. For instance in 1997, only one of these two fields was cultivated. The other lay fallow because of lack of labour to maintain the levees. However, *imoushar* in Banguel are powerful and control a large number of bella. If rice cultivation was a priority for them, they could have ensured enough labour for maintenance. But this would have meant sacrificing a relatively secure livestock production—based on herding—for a far more uncertain agricultural production. Even though the technology and therefore the cultivation system had been improved, through the intervention of the NCA, it was still too risky for people to take the gamble. There was simply not enough labour available in the area to keep livestock as well as to cultivate rice. Therefore, the *imoushar* have continued to prioritize letting the bella herd animals in the vast grazing regions south of the Niger River (Pedersen and Benjaminsen 2008).

Cotton, Food Crops and the Environment in Southern Mali

West Africa is the third most important cotton region in the world after the USA and Uzbekistan, and Mali has long been a leader in the region's cotton production. The Malian cotton zone is situated in the south-eastern corner of the country (Fig. 9.1).

The parastatal cotton company *Compagnie Malienne pour le Développement des Textiles* (CMDT) is a key organization in southern Mali. It is 60% owned by the Malian state and 40% by the French multi-national company GéoCoton. CMDT has a monopoly control over cotton production, processing and marketing in Mali. The parastatal has also regional development as part of its mandate.

CMDT is involved in the whole production chain from research on new cotton varieties, distribution of cotton seeds to the farmers, sale of chemical fertilizers and

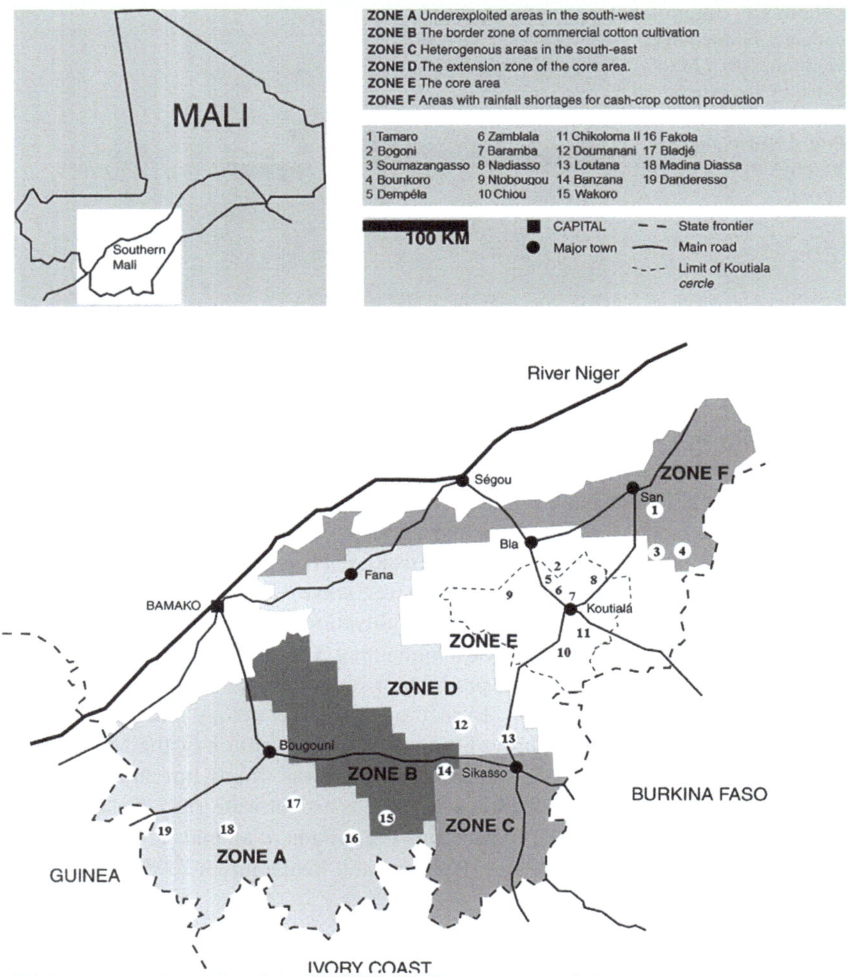

Fig. 9.1 Map of the cotton zone in southern Mali showing the different zones of cotton production intensity and the 19 villages from where soil samples were taken. From Zone E, where there is the most intensive cultivation, to Zone A there is a gradient of decreasing intensity. (Source: Created by the authors)

pesticides on credit, purchase of cotton from farmers' cooperatives, ginning and transport of cotton fibre to international markets. In addition, the company runs social programmes involving adult literacy, education of village cooperatives, land-use planning, soil conservation, as well as other forms of rural development.

Different factors influence the price that farmers will receive for the cotton they grow. First, prices fluctuate internationally, which reflect changes in weather, agricultural policies and demand. In the 1990s, there were relatively high prices on cotton, but from the early 2000s, prices on the world cotton market started to fall. This price

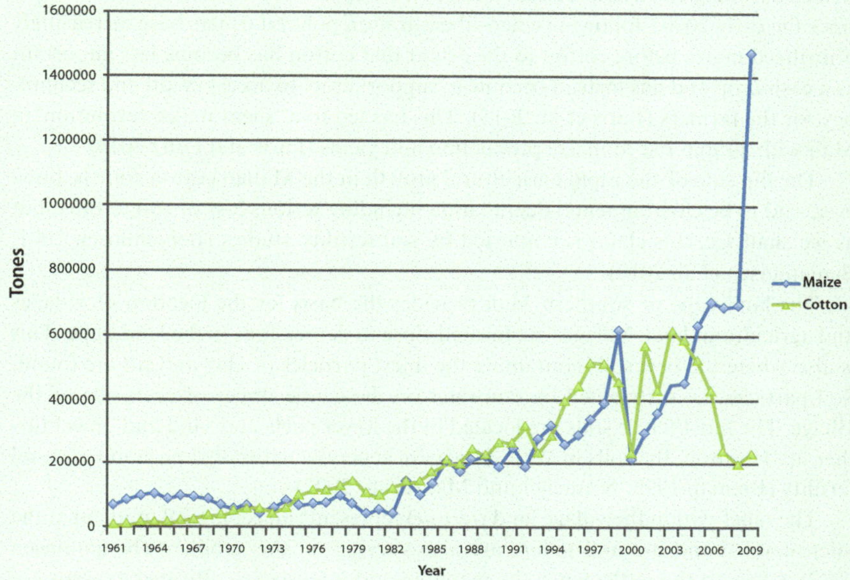

Fig. 9.2 Maize and cotton production in Mali in metric tons 1961–2010. (Source: Laris and Foltz (2014) and FAO)

fall resulted in African critique of the EU and USA for subsidizing their own cotton farmers, and thereby contributing to keep prices as an artificially low level, with serious consequences for a country like Mali.

In addition to reflecting international prices, the annual cotton price in Mali is set before the agricultural season in negotiations between the CMDT and the cotton farmers' union. During the season 2000–2001, the farmers demonstrated that if they are not happy with the price received from the CMDT, they will rather grow food crops like maize or millet. In that single year, cotton production was reduced by 50%.

Cotton production in Mali has, however, often been seen as a success story of African agriculture. This has led cotton to frequently be called 'white gold' and the 'motor of development', and the development of cotton production to be labelled 'Mali's white revolution' (Tefft 2003). The total production of cotton increased from 150 tonnes in 1952 via 3900 tonnes in 1958 to 68,000 tonnes in 1972, 593,000 tonnes in 2003 and 800,000 tons in 2019. While the general trend has been an increase, total production has also been fluctuating during the last couple of decades mainly due to the uncertainty represented by fluctuating international cotton prices (Fig. 9.2).

Food production in the cotton zone has, however, also increased substantially both when cotton production expanded in the 1990s, but especially since the early 2000s. Hence, the cotton zone exports food to other parts of the country and to neighbouring countries. This may partly be an effect of farmers switching between cotton and millet or maize every other year. In this way, food crops will also benefit from the previous year's use of fertilizers. In addition, with unstable cotton prices as well as declining extension services and general support to farmers from the CMDT since the early 2000s, following a neo-liberalization policy, farmers have increasingly prioritized maize before cotton to the extent that cotton has become less important as a cash-crop and has instead become a 'support crop' to access credit and technology for the farmers (Laris et al. 2015). This has led to a 'silent maize revolution' in Mali with a rapid rise in maize production and yields (Laris and Foltz 2014).

The flip side of the rapid agricultural growth in the Malian cotton zone is, however, said to be environmental degradation including serious loss of soil fertility, but as we shall see, this claim is contested by soil fertility studies (Benjaminsen 2001; Benjaminsen et al. 2010).

The landscape of southern Mali provides the basis for the location of villages and agricultural land. Villages are located close to depressions in the landscape. This is also where the best soils containing the finest particles of clay and silt are found. Soil particles are distributed by rainwater on the gentle slopes of each side of the village. The most fertile soils are located in the lower parts and sand and gravel further up. However, the soils in West Africa are generally naturally low in mineral soil fertility (Breman 1998; Niemeijer and Mazzucato 2002).

The relief within the village land (*terroir*) represents an ecological gradient at the micro-level. Land use and tenure are also varying on this gradient (Benjaminsen 2002). Close to the settlements, the manured and intensively cultivated gardens are fenced with branches. Different sorts of vegetables are cultivated here. Most of the products are sold and the one who cultivates also controls the output. In many production units there are also individual fields situated close to the habitations. Outside these gardens and fields, the area with the big fields (*foroba*) of the lineage are located. Normally there is a continuous band of fields, which is almost permanently cultivated. Beyond these big fields, there is 'the bush'. It is often located at the upper parts of the slope, or in general, on more marginal land. Extension of the cultivated area is taking place in 'the bush', since the big fields are more intensively and permanently cultivated. Cotton is usually grown in rotation with cereals such as maize, sorghum and millet in alternate years. The land management model described here, which is common in farming communities in West Africa, involves the transfer of nutrients in the form of manure from the bush to the village fields (Harris 2002).

Until the beginning of the 1980s, fertilizers were subsidized in Mali. This resulted in increased harvests. As part of the World Bank's Structural Adjustment Programme, started in 1982, in Mali the government was, however, forced to phase out these subsidies. Since then cotton production continued to increase, while yields (production per unit area) have declined since the early 1990s (◘ Fig. 9.3).

We argue that the reason why yields decreased in the 1990s into the 2000s is that farmers compensated for the decline in the use of fertilizer per unit area by expanding the cultivated areas. Hence, there was in the 1990s and early 2000s a process of extensification by farmers expanding the area under cotton cultivation and investing less labour and agricultural inputs such as chemical fertilizers per area unit.

In January 1994, the West African Franc (FCFA) was devalued to 50% of its previous value. This increased the profits for cotton growers. While the costs of

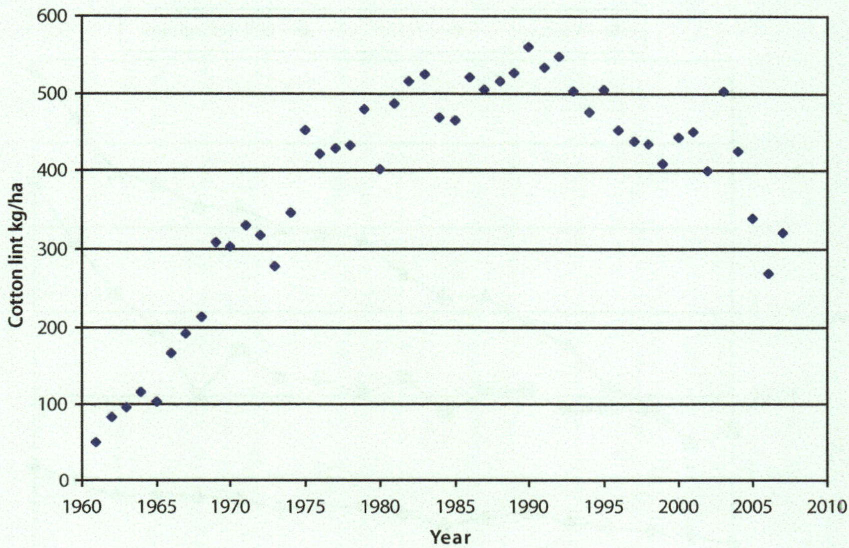

Fig. 9.3 Cotton yields (lint) in Mali. (Source: Benjaminsen et al. (2010) and FAO)

imported farm inputs (e.g. inorganic fertilizers) increased, cotton also became more valuable. In the first years after the devaluation, inorganic fertilizers had never been so cheap, if measured in cotton prices. In addition, in the late 1990s, an agreement between the CMDT and the cotton farmers union reintroduced subsidized fertilizers. This again led to a dramatic rise in fertilizer use. However, the area under cotton cultivation expanded even faster. Figure 9.4 demonstrates the effect of devaluation on area under cotton cultivation in particular for the core area of the cotton zone (Koutiala). Benjaminsen (2001) and Benjaminsen et al. (2010) argue that this explains the fall in cotton yields from 1990.

This argument is, however, contested by Laris and Foltz (2014) and Laris et al. (2015). They hold, also based on a political ecology approach, that the main reason for the fall in yields is that farmers have shifted attention and priority from cotton to maize.

It is, however, possible that both these arguments are correct as the two sets of studies focused on different time periods. The period discussed by Benjaminsen (2001) and Benjaminsen et al. (2010) is mainly the 1990s (with the latter study including a few years after 2000), while Laris et al. (2015) focused on 'the past 15 years'. In the early 2000s, since maize was becoming more profitable than cotton, farmers' priorities clearly shifted from cotton to maize and maize yields boomed, while cotton yields continued to decline (Fig. 9.5).

This agricultural development has necessarily led to considerable changes in the landscape and the environment. Expansion of cultivated land at the expense of pastures and forests represents the clearest and most visible transformation of the landscape. Baramba village, which is centrally located in the cotton zone, can be used to

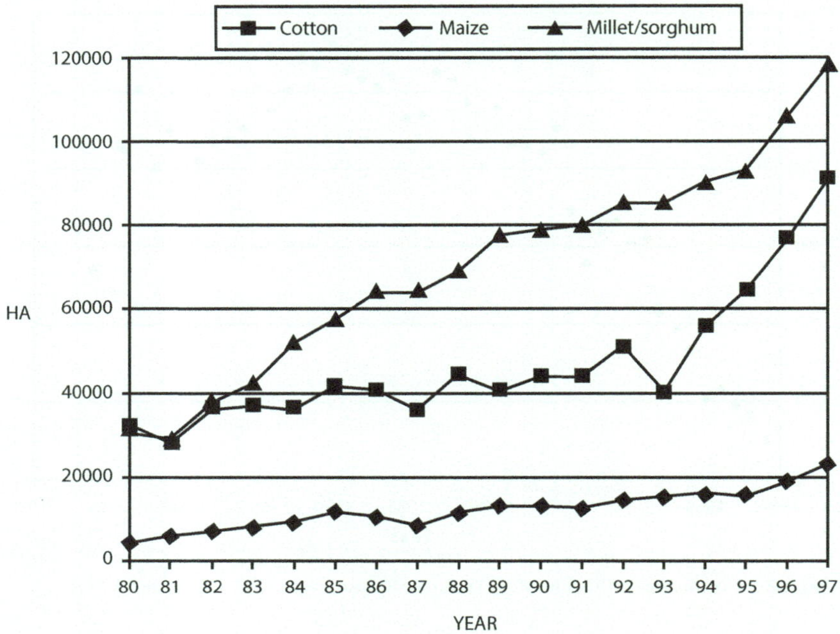

Fig. 9.4 Extension of cultivated area (ha). Main crops. Koutiala district, 1980–97. (Source: Benjaminsen (2001) and CMDT)

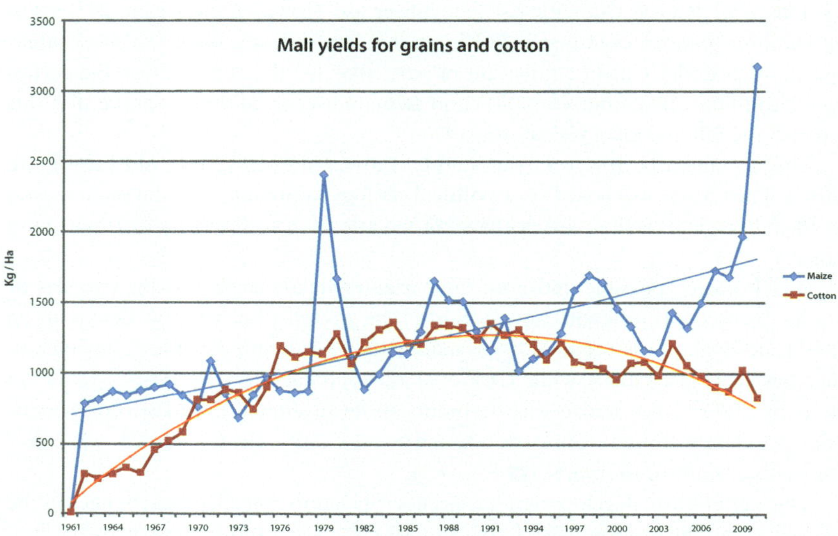

Fig. 9.5 Maize and cotton yields in Mali, 1961–2010. (Source: Laris et al. (2015) and FAO)

9

illustrate the rapid increase in cultivated fields that has taken place since the 1950s. Baramba's village area was mapped using an enlarged aerial photo taken in 1990 (scale 1:15,000) combined with field studies in 1995 (Benjaminsen 2001). Subsequently, a map of land use was established. This was compared to a similar study of aerial photos from 1952 (◘ Fig. 9.6). While cultivated land only covered 12% of the village area in 1952, farmed land increased to 57% in 1990–1995. This increase in farmland at the expense of forests has of course been decisive for the rapid increase in agricultural production that has taken place in this region. However, at the same time the biodiversity in the area has been reduced. Local hunters related in the 1990s that they had to go longer distances to find the game they hunt. What is called 'gallery forests', that is, forests of trees of different layers, have also been considerably reduced in the area.

The loss of soil nutrients ('soil mining') has been presented as a serious environmental threat in the Malian cotton zone. These claims have been based on studies comparing inputs (application of fertilizers and manure, recycling of crop residues, nitrogen fixation, atmospheric deposition of nutrients in rain and dust and enrichment by weathering of soil minerals) and outputs (extraction by crops and losses due to leaching, erosion, volatilization and denitrification) of nutrients. The conclusions have been that there are large nutrient annual deficits of N, K and Mg in the area. For the region as a whole, the deficits were calculated to be 25 kg N/ha, 20 kg K/ha and 5 kg Mg/ha. The cost of this 'soil mining' was said to be equal to 40% of the farmers' total income from agriculture (Van der Pol 1992; Van der Pol and Traoré 1993).

Such nutrient budgets represent a common method of studying soil fertility change. However, it has certain weaknesses as discussed by Scoones and Toulmin (1998, 1999), Ramisch (1999) and Schlecht and Hiernaux (2004). Typically, data for such budgets are collected at the field, farm or village level and extrapolated to whole regions or countries. Results from such nutrient budgets tend to depend on the spatial scale used—the budgets being increasingly negative as the spatial scale increases from the farm to the country level (Schlecht and Hiernaux 2004). Aggregated nutrient budgets therefore tend not to be able to capture local dynamics in time and space.

Another criticism one may raise against studies of nutrient budgets in Mali is related to the quality of the output data of nutrients (erosion, leaching and volatilization). These estimates are not based on measurements in Mali, but rather on secondary data from other countries in West Africa. The uncertainty related to extrapolation of these data to Malian conditions makes it very difficult to give reliable estimates for nutrient budgets in the cotton zone in Mali.

To investigate claims about widespread losses of soil nutrients in the Malian cotton zone, Benjaminsen et al. (2010) used the methods described in the Methods Box in this chapter. The results of the analysis of soil samples showed that the highest content of carbon, nitrogen, phosphorus, potassium and magnesium were, not surprisingly, found in the sacred forests. Beyond these areas, no clear tendency in the relationship between the intensity of cultivation and nutrients in the soils were found. The study did, therefore, not support the idea that intensification of cotton production in Mali results in exhaustion of nutrients in the soil. This is a conclusion that soil studies in more subsistence-oriented production systems in other parts of

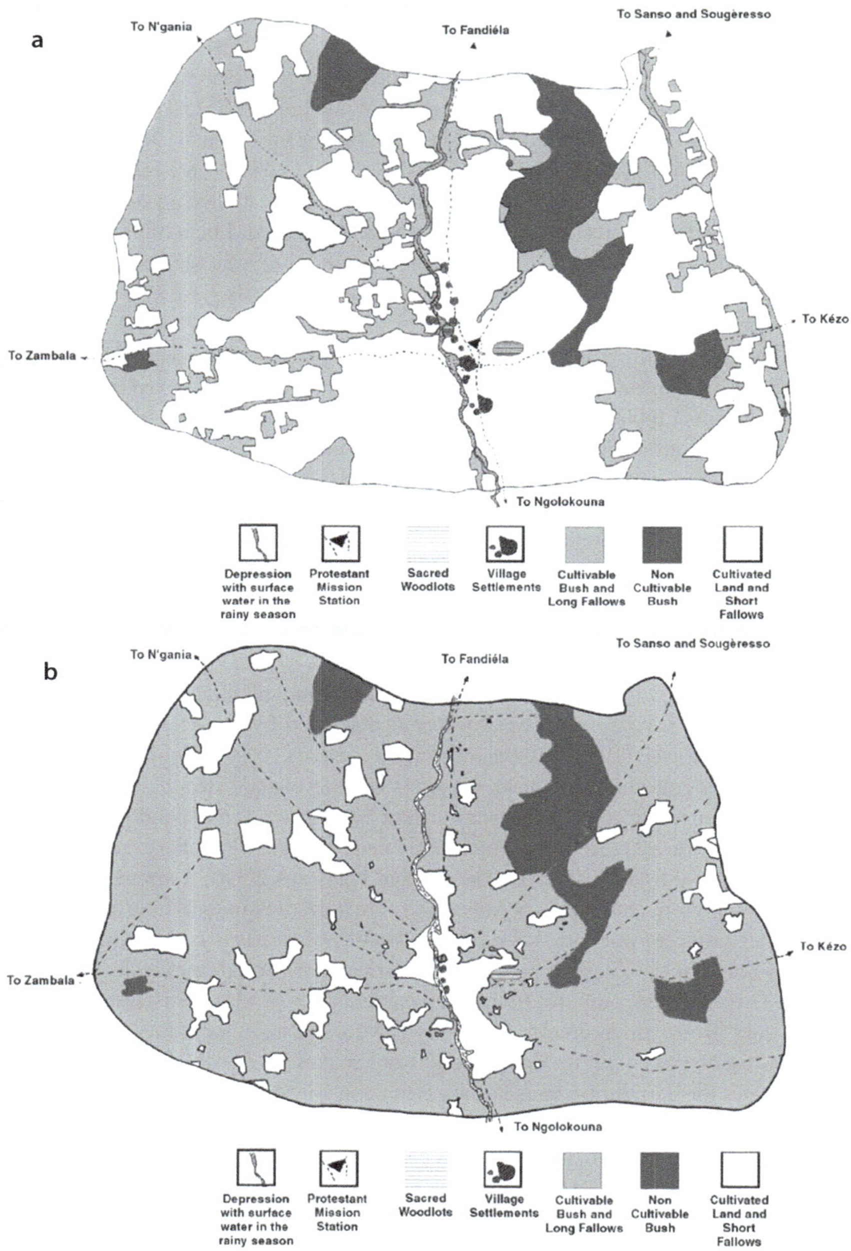

Fig. 9.6 Map of land use changes in Baramba village. The lower map shows the situation in 1952 and the upper map shows land use at the beginning of the 1990s. (Source: Created by the authors)

Africa also have come to (Prudencio 1993; Ramisch 2005; Tittonell et al. 2007).

This example of cotton production in Mali also shows how market access can act as a driving force in agricultural development in line with the study by Tiffen et al. (1994), and it demonstrates that cash-crop cotton cultivation does not necessarily result in exhaustion of soil nutrients.

Marx is sometimes quoted for having said that '... all progress in capitalist agriculture is a progress in the art, not only of robbing the worker, but of robbing the soil' (Marx 1990: 638). Influenced by this idea, it has often been argued in political ecology and development studies more broadly, that the production of commodities for a market necessarily will lead to environmental degradation. This should, however, be treated as a hypothesis to be empirically investigated, rather than as *a priori* fact.

Methods

This box describes the methods used in a study of soil fertility change in the Malian cotton zone by Benjaminsen et al. (2010). Since there was no baseline dataset to compare with, this study had to settle for the spatial analogue method, which consists of comparing soil properties on different sites with varying histories of land use. The main independent variable used was length of fallow. The study did not focus on the effects of cotton production on soil fertility per se, but rather on the varying intensity of the cotton–cereal rotation and how this intensity impacts on soil fertility.

Soil data from cultivated fields were compared with data from fallows of different lengths. In addition, samples were taken on woodland as well as in sacred groves (with permission from the village chief). The woodlands and sacred groves had never been cultivated, but the woodland sites had been grazed. It should also be noted that the sacred groves are usually centrally situated in the villages on some of the best soils. They therefore represent very valuable reference sites being as close to a natural system one can get in terms of soil fertility. The strength of this method is its ability to account for spatial effects of cultivation. Soil fertility is typically higher in the vicinity of the homestead and decreases as the distance from the homestead increases (Prudencio 1993; Ramisch 2005). The soil samples taken from woodlands are distant from the homestead whereas samples taken from cultivated fields and fields under fallow represent soils that are more adjacent to the homestead.

In total, 273 soil samples were collected in 19 villages from north to south in the cotton zone (◘ Fig. 9.1) to cover diversities in agricultural development in the region. The selection was based on the zoning of the cotton zone made by CMDT, where the different sub-zones represent different land use intensities. Following this zoning, the 19 villages were grouped into five different 'gradients', from 1 to 5, with 1 having the highest land use intensity and 5 the lowest.

In addition, soil samples were taken from land that had been continually cultivated for 5 years, land that had been fallow for 5 years, 8–10 years and 15–20 years, forest soil that had never been cultivated and soil from the sacred groves. Many villages in southern Mali have such sacred forests that are fully protected against all human use (farming, grazing and firewood gathering). They are usually located centrally in the villages and on good soil. Therefore, they function as good reference points for the other soil samples.

With the international price increase on food and energy from 2007, land in the Global South, and especially in Africa, was discovered by international capital as an investment object. This again has led a number of journalists, activists, students and scholars to draw critical attention to these investments, and to express a concern for the consequences for smallholders. Small-scale farmers and pastoralists have historically used the areas where many of the investments take place and since they usually do not hold strong formal rights to land and natural resources, they tend to be excluded and dispossessed of land they have used. The term 'land grabbing' was coined to draw attention to this injustice

There is considerable uncertainty regarding how large areas the new land investments cover in total, and the numbers vary a lot. According to the World Bank, 46 million hectares were leased to investors between 2006 and 2010 of which 70% were in Africa. In 2009, FAO estimated, however, that 20 million hectares were leased out in Africa as a result of the new land rush (Evers et al. 2013).

In the debate about these investments, there are two opposing discourses, which again reflect old debates about whether large mechanized units or small family farms best promote agricultural development. On the one hand, there is a modernization discourse arguing that to prevent a Malthusian disaster with overpopulation, agriculture in Africa needs to be modernized through investments that foster technological change, larger units and a closer connection to global markets. Only in this way can total production increase enough to outpace population growth. African agriculture dominated by small family farms is generally seen as unproductive and environmentally destructive. This discourse is supported by most African governments, large investors, many development aid donors and a number of researchers, especially economists.

Definition

In opposition to dominant ideas about the need to modernize African agriculture, there is a discourse about *food sovereignty* where smallholders are seen as the most important land investors, users and producers. Key in this discourse is the idea that the smallholders themselves, and not international capital, should control the management of land, water, seeds and natural resources. While *food security* can be achieved through external support such as food aid, food sovereignty is about giving power to small-scale producers allowing them to control essential parts of the value chain from production to distribution. Only through such empowerment can smallholders achieve increased welfare, is the argument. Moreover, small-scale agriculture is generally seen as sustainable and needs to be strengthened and not marginalized.

This discourse is supported by a number of non-governmental organizations, social and environmental justice activists, as well as many critical scholars including within political ecology. A fair amount of the research supporting this discourse has been published in the *Journal of Peasant Studies*. The international peasant organization *La Via Campesina* (► www.viacampesina.org/en) is also closely connected to this discourse on food sovereignty.

Finally, in this chapter we will discuss two foreign land investments in Tanzania (Bergius et al. 2018). The first is the tree plantations by the Norwegian forest company Green Resources, which started its operations in East Africa in 1997, before the current wave of land investments. The second is Kilombero Plantations Ltd (KPL), which is a large rice plantation owned by the British company Agrica and the Tanzanian parastatal RUBADA. The KPL plantation was established in 2008 as the international interest in African land as an investment object was taking off. Both investments have, at the time of writing, an uncertain economic future.

Green Resources in Tanzania

Green Resources (GR) is Africa's largest forest company with plantations in Mozambique, Tanzania and Uganda. The company, which calls itself 'a leader in carbon finance' owns 45,000 ha of standing forest, claims to employ 3500 people and has 80 shareholders (mostly Norwegian) that have invested about 300 million US dollars in East African plantation forests.

Around 75% of GR's forests are certified by Forest Stewardship Council and there is one plantation in Uganda certified by the Clean Development Mechanism. The company says it plants ten new trees for each tree harvested, and that plantations are only established on 'low value grassland or degraded forestland'. Moreover, the company presents its activities within a sustainable development framework that implies a focus on community development and local benefits. This includes village afforestation, the construction of school buildings, roads and village halls and offices, in addition to providing local employment. In the words of the former CEO Mads Asprem, 'no carbon mitigation activity creates larger economic benefits for the rural poor than afforestation'.

The shareholders of GR have, however, waited for more than 20 years to receive returns on their investments, while the results have been big losses and falling values of shares. For instance, in 2014 the company lost 19.9 million US dollars. This strained financial situation has led to repeated delays in salary payments. Sometimes payments to plantation workers have been delayed by several months. Simultaneously, however, the CEO has enjoyed an annual salary of 401,000 US dollars.

Situated in the Southern Highlands of Tanzania are three of GR's oldest and biggest plantations covering around 74,000 ha. In addition, GR owns East Africa's largest sawmill, Sao Hill, which is also situated in the Southern Highlands. The company and part of the plantations produce timber and the rest are planted to sequester carbon and generate carbon credits. These plantations have been registered under the voluntary carbon standard (VCS) selling credits on the voluntary market.

In total, land has been allocated from six different villages, two villages for each plantation. The villages have not had much to say in this process, as most of the land was negotiated in 1997 before the Village Land Act of 1999, which means that at the time the government and not the village council managed the land. This has led to some of the villages losing more than 33% of their land, which was set as the limit in the Village Land Act (URT 1999) for how much a village can give away to an investor. For example, one of the villages, Uchindile, lost almost 60% of its land to Green Resources (Refseth 2010).

The company has received leasehold for the land for 99 years from the government. According to the Village Land Act, villages cannot lease land directly to investors. The land needs first to be converted to general land, which is managed by the government. This means that when the period for leasehold is out the land is returned to the government and not to the village. In return for giving away land, villages are promised employment, development of infrastructure and support to community projects. In addition, GR has promised 10% of the total revenue from selling carbon credits to the villages (Refseth 2010).

Reports have shown that the benefits from GR projects to the local communities have not been fulfilled as promised (Refseth 2010; Point Carbon and Perspectives 2008; Karumbidza and Menne 2011). Refseth (2010) found for instance that of the promises made in 1997 about one third had been honoured in 2009. At the same time, most workers were only paid 2500 TZS per day (about one US dollar), which is the Tanzanian minimum wage for agricultural work, while the local union claimed that such plantation work should be classified as industrial work, which had a minimum wage of 3000 TZS per day. In addition, work clothes received from the company were not sufficient, most people were employed on short temporary contracts and salary payments often came much too late.

The only roads that had been constructed were roads in the plantation itself, which were not directly benefiting the villages. Despite promises of access to safe water, no efforts to supply water to the villages had been made. Lastly, support to community projects had been slow and barely existed (Refseth 2010). In sum, one main problem with the approach used by GR is that local benefits are not transparently stated in written contracts. Hence, benefits do not become rights that communities hold, but are merely subject to charity from the company.

The claimed climate change mitigation effect from these plantations is also questionable, since vegetation is cleared before the seedlings of pine and eucalyptus are planted. In addition, parts of the plantations in Tanzania have been set on fire by local people several times, which could be seen as a form of resistance against the project. It is obviously also a problem when carbon sold on a market to mitigate

Photo 9.2 Monoculture of pine in a Green Resources plantation. (Photo: Tor A. Benjaminsen)

climate change goes up in smoke in a fire. The effect on biological diversity is clearly also negative when large monocultures of pine and eucalyptus are established (Photo 9.2).

These various consequences will, however, be hidden to buyers of carbon credits from these plantations. There is an information gap between production (tree planting to sequester carbon) and consumption (purchase of carbon credits). This gap is an example of what Marx called 'commodity fetishism' (see ► Chap. 2), which means that the social and environmental consequences of the production are hidden to the consumer. The gap becomes bigger the more global the commodity is and the longer the distance is between production and consumption.

Kilombero Plantations Ltd (KPL) in Tanzania

KPL is a rice plantation of almost 6000 ha located in the fertile Kilombero Valley in Tanzania. In addition to the plantation, KPL also works with surrounding smallholders through an outgrower model based on Sustainable Rice Intensification (SRI). The plantation was established in 2007 as a public-private partnership between Agrica Tanzania Ltd and the parastatal agency Rufiji Basin Development Authority (RUBADA).

Agrica Tanzania Ltd is a subsidiary of UK-based company Agrica Ltd established in 2005 seeking to 'develop sustainable agribusinesses in Africa', while benefitting from the 'compelling investment opportunity' brought about by the increasing global demand for food and feed (Agrica 2011, i). The Norwegian Investment Fund for Developing Countries (Norfund) is one of the major shareholders in Agrica, and its investments amounts to about 10 million US dollars in equity.

In addition to Norfund, a whole range of other development actors have been involved in the project. KPL receives considerable financial and technical support from institutions such as the UK Department for International Development (Dfid), USAID and UNDP among others (Oakland Institute 2015). Moreover, KPL is also partnering with corporate giants such as the Norwegian fertilizer company Yara and the Swiss seed and agro-chemical company Syngenta to strengthen rice value chains in the Kilombero area (Grow Africa 2014).

As with GR, Agrica, through its KPL project, is keen to market itself as a socially responsible investor, which is able to combine profits with caring for communities and the planet. However, critical reports argue that there is a mismatch between this framing of KPL and the experiences of surrounding villagers. These reports show that there are strong contestations around both land and environmental issues, as well as around the outgrower scheme (Oakland Institute 2015; Chachage 2010; Greco 2015).

The area where KPL is located (the Mngeta farm) was originally established in 1986 as a government joint venture between the North Korean and Tanzanian governments (KOTACO). However, after farming only parts of the area KOTACO never became a success and operations ended in 1993. In the period before the establishment of KPL in 2008, local farmers moved in to farm. Some also settled in the area. In addition, pastoralists grazed their livestock on the Mngeta farm. Villagers and pastoralists had, however, also used this land before the arrival of KOTACO. Archival information indicates that the area was cultivated by local farmers as early as the 1920s, while aerial photographs from the 1940s also testify the historical use of this area (Bergius forthcoming). After the arrival of KOTACO in 1986, most of the settled families managed to stay on the land due to the company's limited land-use. Evictions therefore only started in 2007 preparing for KPL's investments (Chachage 2010). To justify these evictions, the company labelled local farmers as 'invaders' and 'squatters' (KPL 2009).

The company agreed, however, to compensate evicted farmers as a 'goodwill gesture' to maintain good relations with local communities. In deciding on the compensation level, KPL claims to have followed World Bank guidelines for social and environmental sustainability. These guidelines promise to improve, or restore, livelihoods of affected persons to levels prevailing prior to the re-establishment of the plantation. However, these guidelines failed to safeguard local interests (Oakland Institute 2015; Bergius forthcoming). Many of the 381 families did not receive any compensation at all, while those who did receive new housing discovered that the compensation houses were built on land that was flooded every rainy season (Oakland Institute 2015; Bergius forthcoming).

In addition, land-use conflicts have increased in the surrounding villages following KPL's investment (Greco 2015). Finally, smallholders surrounding the plantation also complain of crop damage on their farms as a result of the company's spaying of crops by airplane, since sprayed herbicide tended to drift to neighbouring local farms (Oakland Institute 2015). Finally, the replacement of African smallholder farming with British and Norwegian investments to establish a large-scale mechanized farm is a clear example of accumulation by dispossession (see ► Chap. 2). Or, at least the dispossession is very clear, although capital accumulation part has not been very successful, since KPL has been struggling as an economically sustainable business investment leading to economic losses for the investors.

Conclusions

In this chapter, we have argued that a certain population density is necessary for increased agricultural production in Africa. Since investment capital is often lacking in African agriculture, it is mostly people's labour that drives agricultural development. But access to markets is also necessary to create a surplus that can be sold. The case of cotton production in Mali illustrates the potential role of the market as a motor for agricultural development. This, however, requires that smallholders are able to control parts of the commodity chain. In the cotton zone in Mali, farmers hold some power through their union that negotiates a farm-gate price with the parastatal cotton company every year before the agricultural season. If farmers are not satisfied with the price, they will grow more maize and millet, and less cotton. This case also demonstrates that small-scale farming, even when closely linked to a market, does not necessarily lead to loss of soil fertility.

But as the two cases with large-scale land investments in Tanzania show, 'modernisation' through big capital investments in African land-use may also have detrimental effects on sustainability and local livelihoods. There are several problematic aspects with GR's forest plantations such as lack of transparency in the company's social responsibilities, the creation of monocultures of pine or eucalyptus with adverse impacts on biodiversity, and low and delayed salary payments. These impacts remain, however, invisible to buyers of carbon credits leading to this being another case of commodity fetishism. GR has, however, also created a number of local jobs, which should be acknowledged, even if they are poorly paid.

KPL is, however, a clear example of accumulation by dispossession through the establishment of one of East Africa's biggest farms, with British and Norwegian owners, leading to the eviction of 381 families without fair compensation. At least the dispossession part is clear, but the business has struggled economically with losses to the investors and a failed capital accumulation in this case.

Both the GR and KPL investments can also be seen as cases of 'spatial fix', which describes a situation where capital's falling profit rates in any sector, domain or geographical location lead to the migration of capital to non-capitalist spaces and geographical expansion often resulting in processes of accumulation by dispossession.

There is a fair amount of research demonstrating how small-scale farming may be more sustainable than mechanized large-scale farming, and that it can also provide livelihoods for many more people on the same piece of land. Small-scale farmers in Africa are, however, generally fighting an uphill battle against dominating beliefs among governments and development agencies that they should change their 'peasant mentality' and become modern farmers. Large-scale agriculture that these actors tend to promote will necessarily lead to increased landlessness and growth of urban slums. Even if total agricultural production goes up, the food produced will not be available to the ones who in the process lose access to land to feed their families.

Questions

1. Discuss under which conditions market-oriented agriculture can play a positive and negative role for people and environments.
2. How did Ester Boserup see the relationship between population pressure and agricultural production?
3. Explain what is meant by the term 'land grabbing' and why it has been used.
4. Try to identify other examples of commodity fetishism and accumulation by dispossession than those given in this chapter.

References

Agrica. 2011. *Sustainable African agribusiness*. Guernsey and Dar es Salaam: Agrica.

Araghi, F. 1995. Global depeasantization, 1945–1990. *Sociological Quarterly* 36 (2): 337–368.

———. 2000. The great global enclosure of our times: Peasants and the agrarian question at the end of the twentieth century. In *Hungry for profit: The agribusiness threat to farmers, food, and the environment*, ed. F. Magdoff, J.B. Foster, and F.H. Buttel, 145–160. New York: Monthly Review Press.

Benjaminsen, T.A. 2001. The population-agriculture-environment nexus in the Malian cotton zone. *Global Environmental Change* 11 (4): 27–39.

———. 2002. Enclosing the land: Cotton, population growth and tenure in Mali. *Norwegian Journal of Geography* 56: 1–10.

9

Benjaminsen, T.A., J. Aune, and D. Sidibé. 2010. A critical political ecology of cotton and soil fertility in Mali. *Geoforum* 41 (4): 647–656.

Bergius, M. Forthcoming. Rendering history visible: Agribusiness and land acquisitions in Kilombero, Tanzania. *Journal of Eastern African Studies*.

Bergius, M., T.A. Benjaminsen, and M. Widgren. 2018. Green economy, Scandinavian investments and agricultural modernization in Tanzania. *Journal of Peasant Studies* 45 (4): 825–852.

Boserup, E. 1965 (1993). *The conditions of agricultural growth.* London: Earthscan.

Breman, H. 1998. Amélioration de la fertilité des sols en Afrique de l'Ouest: Contraintes et perspectives. In *Soil fertility management in West African land use systems*, ed. G. Renard, A. Neef, K. Becker, and M. von Oppen. Weikersheim: Margraf Verlag.

Brockington, D., O. Howland, V.M. Loiske, M. Mnzava, and C. Noe. 2018. Economic growth, rural assets and prosperity: Exploring the implications of a 20-year record of asset growth in Tanzania. *Journal of Modern African Studies* 56 (2): 217–243.

Chachage, C. 2010. *Land acquisition and accumulation in Tanzania*. Dar es Salaam: PELUM.

Davis, M. 2006. Planet of slums. *New Perspectives Quarterly* 23 (2): 6–11.

Ellis, F. 2000. *Rural livelihoods and diversity in developing countries*. Oxford: Oxford University Press.

Evers, S., C. Seagle, and F. Krijtenburg, eds. 2013. *Africa for sale? Positioning the state, land and society in foreign large-scale land acquisitions in Africa*. Leiden: Brill.

Greco, E. 2015. *Local politics of land and the restructuring of rice farming areas – A comparative study of Tanzania and Uganda*. In The Leverhulme Centre for the Study of Value, School of Environment, Education and Development, The University of Manchester.

Grow Africa. 2014. *Agricultural partnerships take root across Africa – 2nd annual report on private-sector investment in support of country-led transformations in African agriculture*. Grow Africa Secretariat.

Harris, F. 2002. Management of manure in farming systems in semi-arid West Africa. *Experimental Agriculture* 38: 131–148.

Jerven, M. 2013. *Poor numbers. How we are misled by African development statistics and what to do about it*. Ithaca: Cornell University Press.

Karumbidza, B., and W. Menne. 2011. *CDM carbon sink tree plantations: A case study in Tanzania*. Mayville, South Africa: The Timberwatch Coalition.

Ki-Zerbo, J. 1978. *Histoire de l'Afrique noire*. Paris: Hatier.
KPL. 2009. *Farm squatter survey report*. Dar es Salaam: KPL.
Laris, P., and J.D. Foltz. 2014. Cotton as catalyst? The role of shifting fertilizer in Mali's silent maize revolution. *Human Ecology* 42 (6): 857–872.
Laris, P., J.D. Foltz, and B. Voorhees. 2015. Taking from cotton to grow maize: The shifting practices of small-holder farmers in the cotton belt of Mali. *Agricultural Systems* 133: 1–13.
Marx, K. 1990. *Capital: A critique of political economy*. Vol. 1. New York: Penguin.
McKeon, N. 2015. *Food security governance: Empowering communities, regulating corporations*. London and New York: Routledge.
Niemeijer, D., and V. Mazzucato. 2002. Soil degradation in the West African Sahel. How serious is it? *Environment* 44 (2): 20–31.
Oakland Institute. 2015. *Irresponsible investment – Agrica's broken development model in Tanzania*. Oakland: Oakland Institute.
Östberg, W., O. Howland, J. Mduma, and D. Brockington. 2018. Tracing improving livelihoods in rural Africa using local measures of wealth: A case study from central Tanzania, 1991-2016. *Land* 7 (44): 1–26.
Pedersen, J., and T.A. Benjaminsen. 2008. One leg or two? Food security and pastoralism in the northern Sahel. *Human Ecology* 36 (1): 43–57.
Point Carbon and Perspectives. 2008. *CDM due diligence. Idete reforestation project in Tanzania*. Prepared for the Ministry of Finance, Oslo, Norway.
Prudencio, C.Y. 1993. Ring management of soils and crops in the West African semi-arid tropics: The case of the Mossi farming system in Burkina Faso. *Agriculture, Ecosystems and Environment* 47 (3): 237–264.
Radelet, S. 2010. *Emerging Africa: How 17 countries are leading the way*, Center for Global Development, Brief, September 2010. Washington, DC: Center for Global Development.
Ramisch, J. 1999. *In the balance? Evaluating soil nutrient budgets for an agro-pastoral village of Southern Mali*, Managing Africa's soils series no. 9. Edinburgh: IIED Dryland Programme.
———. 2005. Inequality, agro-pastoral exchanges, and soil fertility gradients in southern Mali. *Agriculture, Ecosystems and Environment* 105: 353–372.
Randall, S. 2005. The demographic consequences of conflict, exile and repatriation: A case study of M alian Tuareg. *European Journal of Population* 21: 291–320.
Refseth, T. 2010. *Norwegian carbon plantations in Tanzania: Towards sustainable development?* Master thesis, Norwegian University of Life Sciences.
Schlecht, E., and P. Hiernaux. 2004. Beyond adding up inputs and outputs: Process assessment and upscaling in modelling nutrient flows. *Nutrient Cycling in Agroecosystems* 70: 3003–3319.
Scoones, I., and C. Toulmin. 1998. Soil nutrient balances: What use for policy? *Agriculture, Ecosystems and Environment* 71: 255–267.
———. 1999. *Policies for soil fertility management in Africa*. London: DFID.
Scoones, I., R. Smalley, R. Hall, and D. Tsikata. 2014. *Narratives of scarcity: Understanding the 'global resource grab'*. Cape Town: PLAAS.
Snyder, K.A., E. Sulle, D.A. Massey, A. Petro, P. Qamara, and D. Brockington. 2020. 'Modern' farming and the transformation of livelihoods in rural Tanzania. *Agriculture and Human Values* 37: 33–46.
Tefft, J. 2003. *Mali's white revolution: Smallholder cotton from 1960 to 2003*. Paper presented at the InWEnt, IFPRI, NEPAD, CTA conference 'Successes in African Agriculture', Pretoria, December 1–3, 2003.
Tiffen, M., M. Mortimore, and F. Gichuki. 1994. *More people, less erosion. Environmental recovery in Kenya*. Chichester: John Wiley & Sons Ltd.
Tittonell, P., B. Vanlauwe, N. de Ridder, and K.E. Giller. 2007. Heterogeneity of crop productivity and resource use efficiency within smallholder Kenyan farms: Soil fertility gradients or management of intensity gradients. *Agricultural Systems* 94: 376–390.
United Nations. 2015. *World fertility patterns 2015. Data booklet*. New York: United Nations.
United Republic of Tanzania (URT). 1999. The Village Land Act #5. Dar es Salaam: URT.
Van der Pol, F. 1992. *Soil mining: An unseen contributor to farm income in southern Mali*, Bulletin 325, 48 pp. Amsterdam: Royal Tropical Institute.

Van der Pol, F., and B. Traore. 1993. Soil nutrient depletion by agricultural production in southern Mali. *Fertilizer Research* 36: 79–90.

Wiggins, S. 2005. Success stories from African agriculture: What are the key elements of success? *IDS Bulletin* 36 (2): 17–22.

———. 2014. Presidential address – African agricultural development: Lessons and challenges. *Journal of Agricultural Economics* 65 (3): 529–556.

Woodhouse, P. 2012. *Water in African agronomy, contested agronomy: Agricultural research in a changing world*. London: Routledge.

Stocktake and Ways Forward

Contents

T. A. Benjaminsen, H. Svarstad, *Political Ecology*,
https://doi.org/10.1007/978-3-030-56036-2_10

Overview

We have in this book presented how *we* see this interdisciplinary field called political ecology. In ► Chap. 1, we used the fictional example of the planet Pandora from the movie Avatar to demonstrate conflicts that political ecologists study, and we presented ten syntheses to give a first introduction to what political ecology is all about. In fact, movies, both fiction and documentary, as well as other pieces of art and journalism, can provided cases to be analysed from a political ecological point of view, and we mention some examples of this also in other chapters.

Apart from such examples, the rest of this book is, however, largely drawn from our own empirical research—and from planet Earth. We try to make sense of the various cases by linking them to concepts and theories that are often mobilized within political ecology, most of which are presented in ► Chaps. 1 and 2. There is not much written about how to *do* political ecology. We have therefore included some suggestions in Methods Boxes in some of the chapters. Political ecology is about studying the connections between global, national and local levels of environmental governance, and by using the example of Norway combined with cases from the South in some of the chapters, we have aimed to illustrate how political ecologies in the Global North and South may be interconnected.

10.1 The Approach in This Book

The interconnections between political ecologies in the North and South are illustrated for instance in ► Chap. 6 where we discuss why Norway, with an economy largely dependent on the extraction of fossil fuels, is a world leader in using conservation of tropical forests as a major climate mitigation measure based on the principle of international cost-effectiveness. This principle implies that it allegedly is more cost effective to mitigate climate change in a low-cost country rather than in a high-cost country like Norway. In other words, this is a wealthy nation paying other nations and people to bear the social costs of mitigating climate change, instead of taking action at home. This has been the key principle of Norwegian climate policy since the early 1990s. Recently however, there has been a mounting internal pressure to take more of the cuts in carbon emissions at home rather than abroad, but there is still a lack of concern for climate justice in public debates, in terms of acknowledging the adverse effects of allegedly 'cost-effective' mitigation measures on for instance forest adjacent communities.

In ► Chap. 5, we discuss gender representation in conservation in the North and South through a focus on feminist political ecologies with examples from Norway, Senegal, India and Nepal. Studies in the South often tend to reveal gender inequalities in policy-making on environmental governance. In the North, however, many take it for granted that women have gained a high degree of equity in policy-making in all fields. The chapter shows how rural women in Norway until recently have been denied their legal rights of political representation in processes of citizen involvement in conservation, and we explain this situation through an analysis of power exercised at various levels.

In ► Chap. 7, we show how Norway's policy towards pastoralists cannot be said to be any more favourable to livestock-keepers than what we find in various African countries. There is a large literature demonstrating how African livestock policies are based on misconceptions of pastoral systems, which lead to ill-informed restrictions on these systems and further marginalization of pastoralists including loss of access to key pastures.

This situation is, however, the same in Norway, if not worse, because the modernization pressure on pastoralists is perhaps even more advanced, and the view that pastoralists do not have a place in a modern society is at least as widespread in Norway as it is in Mali or Tanzania. When we have interviewed officials in the Ministry of Food and Agriculture in Norway about Sámi reindeer herding and for instance officials in the Ministry of Agriculture in Tanzania about Maasai pastoralism, we have received more or less identical answers; the pastoralists have too many animals, which leads to overstocking and land degradation, poor economic performance and increasing land-use conflicts. ► Chapters 7 and 8 question such views.

We also show how research, policy-formulation and governance related to Sámi reindeer husbandry are in urgent need of decolonization. The Ministry of Food and Agriculture has for the last 20 years given large funds to ecological research on Sámi reindeer herding, which has been a much-needed source of income for researchers and their institutes. Regardless of what their data actually show, they have presented conclusions stating that there are too many reindeer in proportion to available pastures. This has given a false legitimacy to the continued policy of reducing the number of reindeer and reindeer owners. At the same time, the practical knowledge of the reindeer herders, acquired over generations, is at best briefly mentioned in policy documents, but neglected in practice, and at worst treated as irrelevant or merely as vested in unscientific folklore and superstition.

In ► Chap. 9, we present the example of Norwegian capital investments in forest plantations and large-scale agriculture in Tanzania. The cases demonstrate how 'modernization' through such big capital investments may have detrimental effects on sustainability and local livelihoods. These cases also illustrate some concepts drawn from Marxist political economy such as commodity fetishism, accumulation by dispossession and spatial fix.

Although political ecology has developed in different directions during the last couple of decades, we believe our approach is 'mainstream' in the sense that it is rooted in empirical case studies at the local level relying mostly on qualitative methods, but sometimes also including the analysis of quantitative data. It is based on the idea that one can study the global in the local by first getting a good grasp of what is happening locally meaning what are the ongoing environmental processes and what are their main causes.

When one believes to have a fair understanding of the local situation, the next step would be to move 'upwards' through chains or networks to identify the factors at other levels that influence the local situation. This can be state bureaucracies through laws, policies and discourses and it can be global markets or international actors (e.g. capital investors, big conservation organizations, aid donors). Such a research process may sometimes take political ecologists to new research sites such as capital cities in the Global North (or in new global powers such as India and China) where it may make sense to spend time in order to understand the root causes of the processes studied. Often these causes are historical, and many political ecologists have therefore, through following chains or networks, decided to spend time in colonial archives.

▶ Chapter 4 discusses the establishment and governance of protected areas in the Global South including examples from Tanzania, South Africa and India. The governance of protected areas that we observe today is deeply rooted in practices that can be traced back to the colonial period. In addition, current governance is strongly influenced by powerful international actors such as big conservation organizations and aid donors. These actors are not only powerful because of the funding they can provide, but also because they have discursive power by being able to dominate international discourse production on the conservation of for instance elephants or tigers. They have been highly successful in producing and reproducing a global win-win discourse on conservation where the protection of wildlife and ecosystems is said to also provide communities with income and well-being.

In contrast, political ecologists have abundantly demonstrated, through a large number of case studies, how conservation practice often leads to exclusion and dispossession of local communities. In addition, the violence performed by anti-poaching units and park rangers has increasingly been a theme in the literature on

the political ecology of conservation. The negative effects of much conservation practice on livelihoods as well the violence and human rights abuses sometimes involved have also been documented by journalists in documentary films and articles. A recent example is a series of articles in BuzzFeed News about how WWF has funded park rangers who are responsible for serious human rights abuses including murder and rape (see 'WWF's Secret War' on ▶ https://www.buzzfeednews.com/collection/wwfsecretwar).

Such journalism in addition to the scholarly work by political ecologists challenge the dominating win-win discourse on conservation and demonstrate the need for increased environmental justice in conservation practice. In ▶ Chap. 3, we introduce ways to analyse discourses on particular themes and connected narratives, and to better understand the exercise of discursive power and its resistance.

The key themes of power, justice and environmental sustainability run through all chapters in this book. Our approach to political ecology is also mainstream in the sense that it mainly combines discourse and narrative analysis with Marxist political economy. In addition, we draw on insights from natural science when this is relevant.

10.2 Where Do We Go from Here?

As a dynamic field in steady development, political ecology has during the last few decades continued to add new theoretical approaches and themes to its repertoire. Here, at the end of this book, we will briefly mention three recent developments that we think will play an increasing role in this field in the years to come. These developments also to some extent represent challenges to and potential critiques of political ecology.

We highlight these three directions in particular because they are relatively new developments and because we personally find them to be promising trends in the field. These are (1) the return to discussions about limits through links to the degrowth movement, (2) the interaction between political ecology and environmental justice, and (3) decolonization, and what this debate means for political ecology.

10.2.1 Degrowth

One of the first contributions to political ecology and therefore one of the longest traditions in the, after all, relatively short history of this field has been the critique of Malthusianism and deconstructions of notions of scarcity, limits and carrying capacity. A number of political ecology studies, including some presented in this book, have shown how such ideas tend to serve elites and powerful actors, in addition to often being based on poor science. Political ecology studies have documented how narratives about degradation and scarcity have served to justify elite capture, marginalization and dispossession including forms of accumulation by dispossession, and to further contribute to exacerbate land-use conflicts.

The imminent climate crisis with its potential planetary boundaries and tipping points and the logical contradiction between continued economic growth and a confined atmosphere, may however, increasingly represent a challenge to this traditional political ecology position. Some may argue that there is no contradiction here and that this apparent conflict is more a question of scale. But still the debate should be taken and this question needs to be confronted head on in a traditional political ecology way. The debate between Robbins (2020) and Gómez-Baggethun (2020) and the book by Kallis (2019) were kick-off contributions (see ▸ Chap. 2) to what we expect will be a wide and perhaps intense debate in the years to come.

10.2.2 Environmental Justice

Environmental Justice and political ecology are related fields with many parallels in their common focus on justice and environmental sustainability (▸ Chap. 2). But surprisingly, there have until recently been few interactions and direct engagements between these two approaches to critical environmental studies. Environmental Justice has developed in different directions with a wide range of contributions. In terms of number of publications, this is also a much larger literature than the one in political ecology, and it represents a literature that political ecology scholars would benefit from engaging more with. There is potentially a lot to learn from Environmental Justice and its systematic treatment of notions of justice (see Svarstad and Benjaminsen 2020).

10.2.3 Decolonization

When students at the University of Cape Town managed to get the statue of Cecile Rhodes removed from its viewpoint over the city close to the university in April 2015, this event started a global debate about decolonization in academia.

This debate has focused on questions such as: Whose worldviews dominate in academia? Whose values are prioritized? How can perspectives from indigenous and colonized peoples be included in research and teaching? And how does all this matter for how environments and natural resources are governed?

This latter question is, in particular, important in political ecology, and this is also a question that political ecologists have spent some time studying and discussing. In fact, political ecologists have been at the forefront to decolonize environmental science (Neimark et al. 2019), as well as related themes such as 'conservation' (Adams and Mulligan 2003; Martin et al. 2016; Salazar Parrenas 2018), 'development' (Wainwright 2008), 'environmental education' (Meek and Lloro-Bidart 2017) and 'food justice' (Bradley and Herrera 2016).

We would argue that the unique contribution of political ecology to the decolonization debate will be on how (neo)colonial power relations lead to indigenous and other marginalized people losing access to land and resources. This struggle for land is closely linked to a struggle over meaning where some research feed into dominating policy narratives that serve to justify dispossession. Hence, in such cases, decolonization would mean changing the narratives through providing better empirical research.

Finally, this means that decolonization does not necessarily need to go via postcolonial or poststructural theories. A critical investigation into the empirical basis for colonial narratives may provide a foundation for an argument about decolonization, as shown in the case about Sámi reindeer pastoralism in ▶ Chap. 7. Hence, decolonization is not in contradiction to science, as some seem to think—only to poor science.

References

Adams, W., and M. Mulligan, eds. 2003. *Decolonizing nature: Strategies for conservation in a post-colonial era*. London: Earthscan.

Bradley, K., and H. Herrera. 2016. Decolonizing food justice: Naming, resisting, and researching colonizing forces in the movement. *Antipode* 48 (1): 97–114.

Gómez-Baggethun, E. 2020. More is more: Scaling political ecology within limits. *Political Geography* 76: 102095.

Kallis, G. 2019. *Limits: Why Malthus was wrong and why environmentalists should care*. Palo Alto: Stanford University Press.

Martin, A., B. Coolsaet, E. Corbera, N.M. Dawson, J.A. Fraser, I. Lehmann, and I. Rodriguez. 2016. Justice and conservation: The need to incorporate recognition. *Biological Conservation* 197: 254–261.

Meek, D., and T. Lloro-Bidart. 2017. Introduction: Synthesizing a political ecology of education. *Journal of Environmental Education* 48 (4): 213–225.

Neimark, B., J. Childs, A.J. Nightingale, C.J. Cavanagh, S. Sullivan, T.A. Benjaminsen, S. Batterbury, S. Koot, and W. Harcourt. 2019. Speaking power to 'post-truth': Critical political ecology and the new authoritarianism. *Annals of the American Association of Geographers* 109 (2): 613–623.

Robbins, P. 2020. Is less more … or is more less? Scaling the political ecologies of the future. *Political Geography* 76: 102018.

Salazar Parrenas, J. 2018. *Decolonizing extinction. The work of care in orangutan rehabilitation*. Durham: Duke University Press.

Svarstad, H., and T.A. Benjaminsen. 2020. Reading radical environmental justice through a political ecology lens. *Geoforum* 108: 1–11.

Wainwright, J. 2008. *Decolonizing development: Colonial power and the Maya*. Oxford: Blackwell.

Supplementary Information

T. A. Benjaminsen, H. Svarstad, *Political Ecology*,
https://doi.org/10.1007/978-3-030-56036-2

Index

D

E

F

N

O

P

Q

R

S

T

U

V

W

Y

Z

The manufacturer's authorised representative in the EU is Springer Nature Customer Service Centre GmbH, Europaplatz 3, 69115 Heidelberg, Germany. If you have any concerns regarding our products, please contact ProductSafety@springernature.com

Printed and bound by CPI Group (UK) Ltd, Croydon, CR0 4YY
26/08/2026
02204378-0007